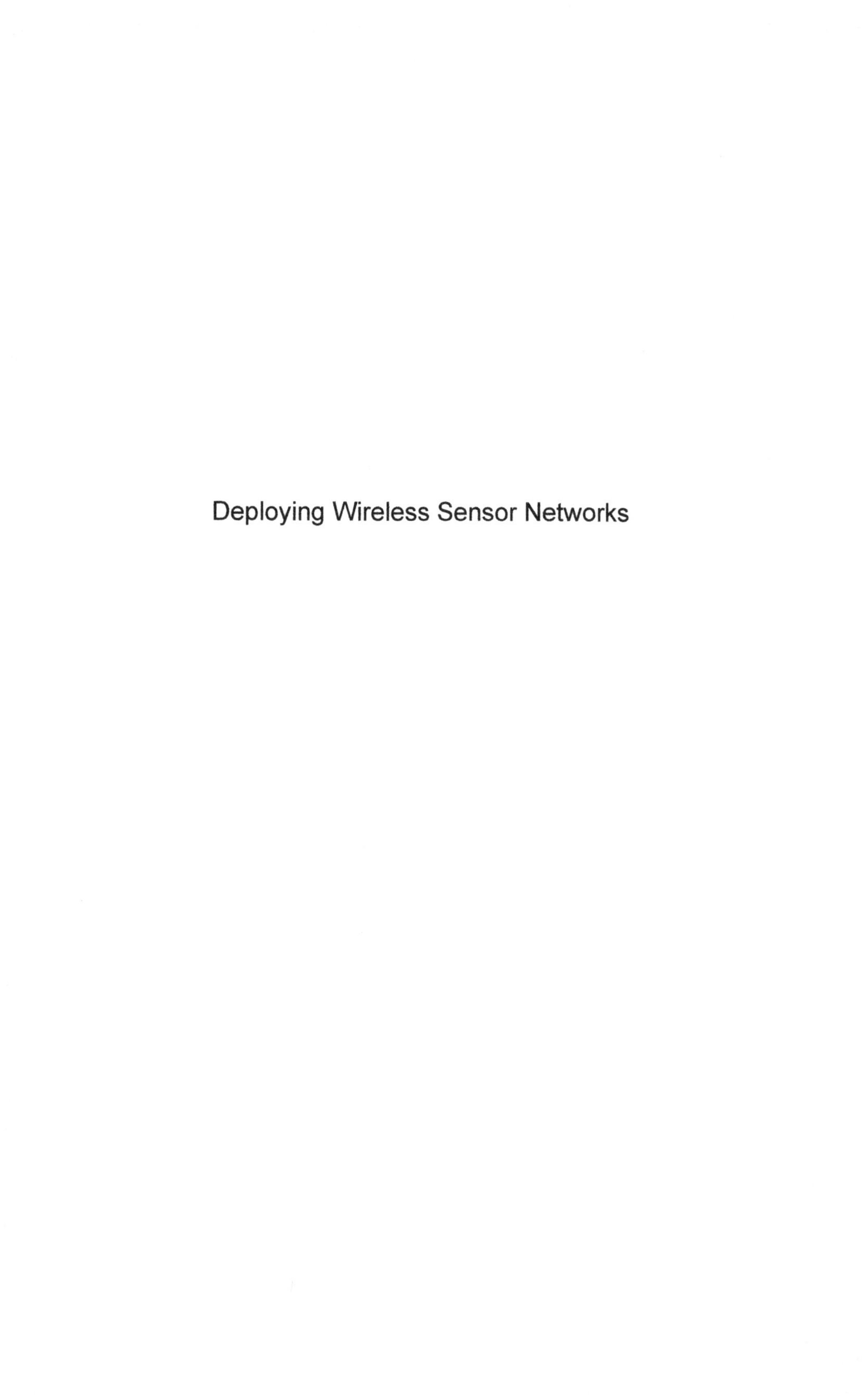

Deploying Wireless Sensor Networks

Sensor Networks Set

coordinated by
Abdelhamid Mellouk

Deploying Wireless Sensor Networks

Theory and Practice

Mustapha Reda Senouci
Abdelhamid Mellouk

First published 2016 in Great Britain and the United States by ISTE Press Ltd and Elsevier Ltd

ISTE Press Ltd
27-37 St George's Road
London SW19 4EU
UK

www.iste.co.uk

Elsevier Ltd
The Boulevard, Langford Lane
Kidlington, Oxford, OX5 1GB
UK

www.elsevier.com

British Library Cataloguing-in-Publication Data
A CIP record for this book is available from the British Library
Library of Congress Cataloging in Publication Data
A catalog record for this book is available from the Library of Congress
ISBN 978-1-78548-099-7

Printed and bound in the UK and US

Contents

Preface

The last two decades of the 20th Century have been driven by the emergence and evolution of many types of networks and there is no doubt that in the 21st Century the concept of the Internet of Things (IoT) and its applications will have a key role in the way we understand our society. The IoT describes the trend for environments, buildings, vehicles, clothing, portable devices and other objects to have a digital representation and the ability to sense, use or exchange information. IoT makes things more interesting by connecting real-world objects, places and people through the digital world. Small objects connected through the IoT are today considered to be one of the main challenges for the business revolution in the coming years. A widespread use of such connected objects will undoubtedly influence people, societies and businesses.

In recent years, the continuous evolution of technologies and the development of new applications and services have steered networking research toward new problems, which have emerged as the network evolves with new features. This has moved toward what is usually referred to as the Future Internet which has become one of the basic infrastructures that currently supports the global economy. In fact, there is a strong need to build a new network scenario, where networked computer devices are proliferating rapidly, supporting new types of services, usages and applications: from wireless sensor networks and new optical network technologies to cloud computing, high-end mobile devices supporting high definition media, high performance computers, peer-to-peer networks and various platforms and applications. This new network scenario is fueling research in the area of new network architectures which consider both the requirements and demands of key emerging applications and services and the currently deployed network infrastructures.

One of the most promising technologies within the IoT today consists of sensor networks. It will be fascinating to look back in the years ahead and note the growing interest in the use of sensor networks in real-world applications. In fact, these kinds of networks represent an efficient technology to monitor and collect specific characteristics in any environment. Several applications have already been envisioned, in a wide range of areas such as military, commercial, emergency, biology and health care applications. A sensor is a physical component able to accomplish three tasks: identify a physical quantity, treat any such information and transmit this information to a sink. In most practical applications, sensors do not change their locations once they are deployed on the sensing field. One of the critical and important aspects in the success of the use of these networks in real-life applications is correlated to the deployment of sensors.

Due the emergence of different kinds of sensors and the foreseen proliferation of different and specific types of services supported by these sensors, the use of networks based on a large amount of sensors has the potential to become a real challenge when taking several kinds of applications into account.

The book focuses on the current state-of-the-art research results and experience reports in the area of deployment techniques dedicated to wireless sensor networks (WSNs). It details all the deployment approaches used in this area in the case of static and mobile environments. Moreover, an in-depth discussion concerning deployment-related issues such as deployment cost, coverage, sensors uncertainty, connectivity, sensors reliability, network lifetime and harsh deployment environments is provided. This book shows that WSN deployment field is a very dynamic area in terms of theory and application.

To give a complete bibliography and a historical account of the research that led to the present form of the subject would have been impossible. It is thus inevitable that some topics have been treated in less detail than others. The choices made reflect in part personal taste and expertise, and in part a preference for a very promising research and recent developments in the field of sensor deployment techniques.

This book is a start, but also leaves many questions unanswered. We hope that it will inspire a new generation of researchers.

The authors hope you will enjoy reading this book and hope to provide the readers with many helpful ideas and overviews for your own study.

Mustapha Reda SENOUCI
Abdelhamid MELLOUK
February 2016

Introduction

Recent years have witnessed successful real-world deployments of wireless sensor networks (WSNs) in a wide range of civilian and military applications. In most practical applications, sensors do not change their locations once they are deployed in the sensing field. We call this kind of deployment practice *static sensor deployment*. In the literature, static sensor deployment is typically carried out in one of two approaches: *random deployment* or *deterministic deployment*. The selection of a suitable approach depends on many factors, such as the type of sensors, the nature of the region of interest (RoI), and the application needs. However, when sensors are able to move on their own, dynamic deployment reconfiguration can be exploited to enhance the network performance. We call this kind of deployment practice *dynamic sensor deployment*.

This book addresses WSN deployment, which is a mandatory and critical step in the process of developing WSN solutions for real-life applications. The discussion starts with simple approaches to deploy static WSNs and then is extended to more sophisticated approaches to deploy mobile WSNs. Moreover, an in-depth discussion concerning deployment-related issues such as deployment cost, coverage, sensors uncertainty, connectivity, sensors reliability, network lifetime and harsh deployment environments is provided.

This book is divided into five chapters. After introducing WSNs (Chapter 1), the book provides an in-depth investigation of WSN deployment approaches that generally fall under one of the following categories: random deployment (Chapter 2), deterministic deployment (Chapter 3), fusion-based deterministic deployment (Chapter 4) and dynamic deployment (Chapter 5).

Chapter 1 provides an overview of WSNs. It highlights the major characteristics of sensors and WSNs, followed by an introduction to typical WSN models and applications. Further, this chapter explains the research problems explored in this book, i.e. WSN deployment issues and their importance to WSN applications.

Chapter 2 focuses on the most *naive* approach to deploy static WSNs: random deployment. Due to the large scale of WSNs or to the inaccessibility/harshness of the RoI, random deployment is often the best choice. Sensors may be deployed from a plane, delivered in an artillery shell, rocket or missile, or catapulted from a shipboard. Such random sensor placement strategies are discussed in Chapter 2.

Although the ease and practicalities of random deployment are appealing, it is often considered too expensive in comparison to deterministic deployment. The latter is optimal as sensors are placed at predetermined coordinates to guarantee network efficiency. Chapter 3 investigates static WSN deterministic deployment, which has different appellations in the literature, e.g. placement, layout, coverage, or positioning problems in WSNs. It highlights the components involved and discusses the existing literature. Moreover, it analyzes the uncertainty-aware WSN deployment where sensors may not always provide reliable information, and shows how evidence theory could be used to design better deployment strategies. A comprehensive methodology for deterministic deployment of WSNs is presented and executed to deploy a simplified indoor surveillance WSN for motion detection.

Chapter 4 investigates the fusion-based deterministic deployment that is usually employed in the deployment of WSNs for critical applications that impose stringent requirements such as a high detection rate coupled with a low false alarm rate. This chapter discusses existing sensor placement algorithms and shows how evidence theory could be exploited to design better fusion-based deployment strategies. As an example, this chapter reports the obtained results when deploying a simplified fusion-based indoor surveillance WSN.

As mentioned previously, WSNs can be formed by dropping sensors from the air. However, random deployment of sensors can leave holes in terms of coverage in the RoI. Sensors mobility could be exploited to improve the random initial deployment. Such a practice is called movement-assisted sensor deployment, or WSN self-deployment. Moreover, sensor failure may cause connectivity loss, and in some cases, network partitioning. Dynamically repositioning the sensors while the network is operational is necessary to deal with such events. In other words, sensor relocation could be used to provide WSNs with self-healing capabilities. Chapter 5 reviews recent literature pertaining to WSN self-deployment and self-healing strategies. Classifications of the most recent deployment techniques are provided. Moreover, different proposed algorithms are categorized, summarized and compared.

At the end of each chapter, practical issues that need further research are discussed. In summary, this book discusses both theoretical and practical aspects and provides guidelines for effective deployment of WSNs.

1

Wireless Sensor Networks

Recent technological advances in the field of microelectromechanical systems (MEMS) have enabled the development of low-cost, low-power and small-scale sensor nodes that integrate sensing, processing, storage, and communication capabilities. Such sensors may be deployed in large numbers over vast geographical areas to form a Wireless Sensor Network (WSN), which provide unprecedented opportunities for monitoring and controlling the real world. This chapter provides a brief overview of WSNs, starting with a general definition of WSNs followed by an introduction to typical WSNs models and applications. Then, this chapter explains the research problems explored in this book, i.e. WSNs deployment problems, and their importance to WSNs applications.

1.1. WSN definition, models and applications

1.1.1. *Sensor nodes*

Commonly, a sensor is a device that responds to physical quantities such as heat, and converts them into electricity to enable automatic interpretation and processing. A sensor node (generally referred to as sensor or mote) is an autonomous, compact device that not only integrates sensors but also includes other units to process and deliver sensory data. Thus, a mote generates data from sensing physical parameters, and eventually transmits this data to a main location. A typical sensor node comprises the following units: sensor, communication, microcontroller, memory and power. Depending on the application requirements, other units could be included such as: GPS, locomotory, energy harvesting, etc.

A sensor node is implemented by a sensor board that integrates all the aforementioned components and other necessary circuitry. Figure 1.1 presents a photograph of an open source hardware Arduino UNO-based sensor node. The microcontroller is based on the Atmel AVR ATmega328 with 32 KB of program memory, 2 KB of data memory, and 1 KB of EEPROM. It can operate from a wide range of power-supply voltages, from 1.8 V to 5.5 V. This makes it well suited for battery-powered applications.

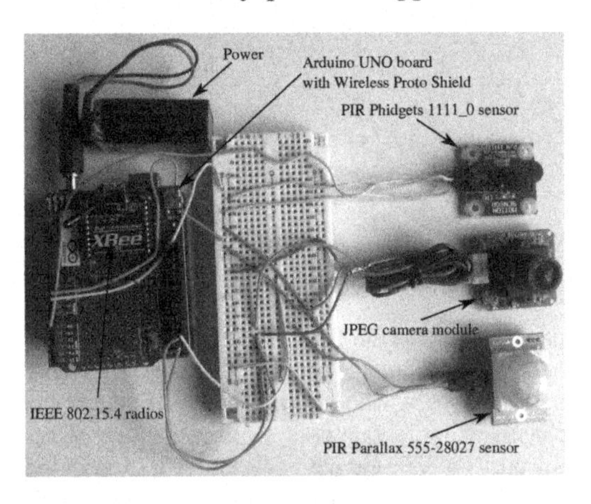

Figure 1.1. *Photograph of an Arduino-based sensor node. For a color version of the figure, see www.iste.co.uk/senouci/wireless.zip*

The Arduino UNO has 14 digital input/output pins and six analog input pins. The architecture of the Arduino board exposes these pins so they can be connected to external circuits easily. The sensor board can support many sensors types and communications modules. In Figure 1.1, the sensor node includes a digital motion detector (PIR Parallax 555-28027 sensor), an analogue motion detector (PIR Phidgets 1111_0 sensor), a camera module, an IEEE 802.15.4 radio module (XBee module), and a battery. It should be noted that the sensor node presented above is more educational than professional, and it will be used in the next chapters to deploy a surveillance WSN. For professional applications, besides application-specific solutions, high-performance sensor nodes such as Stargate and Imote2 [MOO 10] might be employed. We will discuss various WSNs architectures and models. Before that, we first introduce the concept of sensor coverage and sensor communication models in the following sections.

1.1.2. *Sensor coverage models*

From a mathematical point of view, a sensor coverage model is a function that accepts input from parameters such as the distances (and the angles) between the sensors and a point in space. The output of such a function is the coverage measure. The analysis of sensor coverage models shows the existence of different classifications based on distinct goals [WAN 10a]. In this book, these models are classified into three groups: (i) binary coverage models; (ii) probabilistic coverage models, and (iii) evidential coverage models. A taxonomy for sensor coverage models is depicted in Figure 1.2. In what follows, we present some commonly used coverage models in detail.

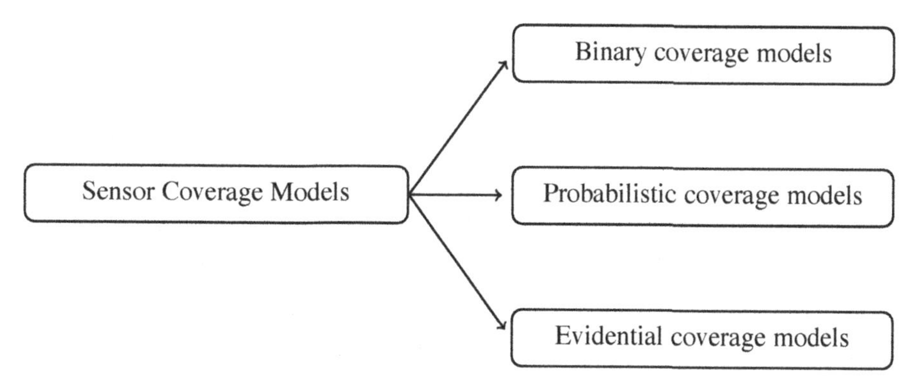

Figure 1.2. *A taxonomy for sensor coverage models*

Because of its simplicity, the binary coverage model has been widely used [WAN 10a]. In this model, sensors are modeled as having a predetermined range of effectiveness. Coverage within the range, which is typically characterized by a disk (but can also be any arbitrary shape or a collection of shapes), is assumed to be effective and coverage outside of the given range is assumed to be non-effective. A well-known variant of the binary model is the disk model, wherein the sensing area of a sensor is often modeled as a disk with radius R_s (sensing range) centered at the sensor's location. For an event that occurs at p, the following equation calculates the probability of detection of that event by a sensor s:

$$P_{s/p} = \begin{cases} 1 & \text{if } \|sp\| \le R_s \\ 0 & \text{otherwise} \end{cases} \qquad [1.1]$$

where $\|sp\|$ is the Euclidean distance between s and p.

When using the binary model, several WSN problems (e.g. the coverage problem) are mapped to geometric problems, which simplify the analysis. However, the binary model does not consider the stochastic nature of sensing which could cause erroneous estimation of system performance in the real world. For that, many recent research works use *probabilistic coverage models* to capture the stochastic nature of sensing. An example of a probabilistic coverage model is given by:

$$P_{s/p} = \frac{C}{\|sp\|^{\gamma}} \qquad [1.2]$$

where C is a constant, and γ is the path attenuation exponent.

The coverage measure is inversely proportional to the point-sensor distance. When this latter becomes very large (resp. very small), the coverage measure might be assumed null (resp. full). Examples of such practices are truncated probabilistic coverage models such as the following model [ZOU 05]:

$$P_{s/p} = \begin{cases} Ce^{-\delta\|sp\|} & \text{if } \|sp\| \leq R_s \\ 0 & \text{otherwise} \end{cases} \qquad [1.3]$$

where δ is a parameter representing the characteristics of the sensor. Another model [ZOU 05] is defined as follows:

$$P_{s/p} = \begin{cases} 1 & \text{if } \|sp\| \leq R_s - R_u \\ e^{-\alpha(\|sp\|-(R_s-R_u))^{\beta}} & \text{if } R_s - R_u < \|sp\| \leq R_s \\ 0 & \text{if } R_s < \|sp\| \end{cases} \qquad [1.4]$$

where α and β are constants, and R_u is called the uncertain range. Figure 1.3 shows the coverage measure as a function of the sensor-point distance for the above-mentioned coverage models.

Recently, an evidence-based sensor coverage model based on the transferable belief model has been proposed in [SEN 12c]. This coverage model not only considers the imperfections associated with sensor readings, but can be easily extended to include deployment-related issues, such as sensor reliability.

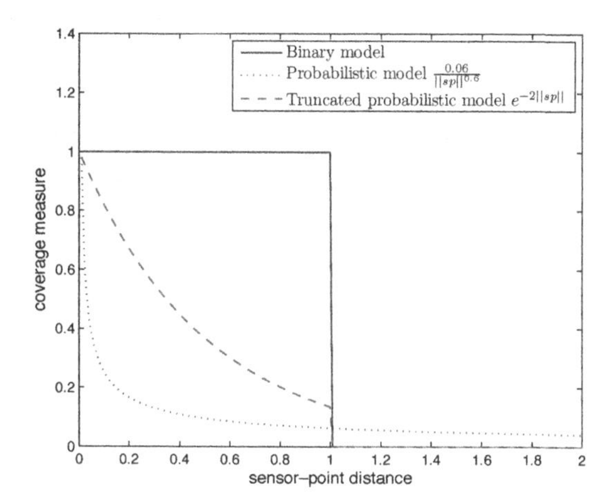

Figure 1.3. *Coverage measure vs. the sensor-point distance for the binary model, probabilistic model, and truncated probabilistic model. For a color version of the figure, see www.iste.co.uk/senouci/wireless.zip*

In the evidence-based coverage model [SEN 12c], two states are required to specify whether a space point $p \in RoI$ is covered: θ_0 (*not covered*) and θ_1 (*covered*). Thus, the frame of discernment (FoD) is the set $\Theta = \{\theta_0, \theta_1\}$. Each sensor s provides information on the coverage of a space point $p \in RoI$ with a belief $x_{s/p}$. The complementary information $1 - x_{s/p}$ is assigned to the whole FoD because it encodes the sensor ignorance. The output from the sensor s about a space point $p \in RoI$ can thus be represented as a basic belief assignment (*bba*) $m_{s/p}$, with two focal sets: the singleton $\{\theta_1\}$ and the FoD Θ, defined as follows:

$$
\begin{aligned}
& m_{s/p}(\{\theta_1\}) = x_{s/p},\ x_{s/p} \in [0, 1] \\
& m_{s/p}(\Theta) = 1 - x_{s/p}, \\
& m_{s/p}(\emptyset) = 0
\end{aligned}
\qquad [1.5]
$$

Relatively to a space point p, a sensor s provides $m_{s/p}$ as a belief function. To decide whether p is covered by s, the pignistic transformation of $m_{s/p}$ (denoted by $BetP_{s/p}$) is constructed. The decision is based on selecting the hypothesis $\hat{\theta}$ with the largest pignistic probability:

$$\hat{\theta} = \arg\max_{i=0,1} BetP_{s/p}(\{\theta_i\}).$$

The sensor coverage model characterizes the observability of physical phenomenas by an individual sensor. On the other hand, *network coverage* can be perceived as a consensus measure delivered by a network of distributed sensors. The network coverage at point p, denoted $\mathbb{P}_p$, is estimated as:

$$\mathbb{P}_p = 1 - \prod_{s \in RoI} (1 - P_{s/p}) \qquad [1.6]$$

where $P_{s/p}$ is the probability that a sensor s detects an event at p (equations [1.2]–[1.4]). In the case of the evidence-based coverage model [SEN 12c], for N sensors, the combination of the N *bbas* $m_{1/p}, \ldots, m_{N/p}$ using the conjunctive rule yields a *bba* m_p with 2 focal sets: $\{\theta_1\}$ and the FoD Θ. This *bba* has the following expression:

$$m_p(\{\theta_1\}) = \prod_{i=1}^{N} x_{i/p} + \underbrace{x_{j/p} x_{k/p} \ldots x_{L/p}}_{\substack{1:N-1 \;\; terms \\ j,k,\ldots,L=1\ldots N \\ j \neq k \neq \ldots \neq L \neq i}} \sum_{i=1}^{N} (1 - x_{i/p})$$

$$m_p(\Theta) = \prod_{i=1}^{N} (1 - x_{i/p})$$

1.1.3. *Sensor communication models*

Wireless sensor nodes communicate via their radio modules. Two nodes are directly connected if they can transmit/receive data to/from each other. A sensor communication model (or a transmission model) is a mathematical model that quantifies the direct connectivity between sensor nodes.

A commonly assumed communication model is the disk connectivity model according to which a sensor node can communicate with other nodes located within a disk itself centered within the radius of its communication range R_c. In other words, two sensors are able to communicate directly if they are within one communication hop of each other. This model considers network connectivity mainly from a geometric perspective, which simplifies the analysis. However, it remains limited and unrealistic. Indeed, empirical studies [NIK 93, SEN 14a] show that there is no clear cut-off boundary between successful and unsuccessful communication.

In practice, the attenuation experienced by a wireless signal at a given distance is described by the path loss, whereas shadowing describes random fluctuations in signal strength at a known path loss. Empirical measurements have indicated that shadowing is a zero-mean normally distributed random variable with standard deviation σ_ϵ. Due to the unique characteristics of each environment, most radio propagation models use a combination of analytical and empirical methods. One of the most common radio propagation models is the log-normal shadowing path loss model [RAP 01] which is given by:

$$PL(d) = PL(d_0) + 10\gamma \log_{10}(\frac{d}{d_0}) + \epsilon \qquad [1.7]$$

where d is the transmitter–receiver distance, d_0 a reference distance, γ the path loss exponent (rate at which signal decays), and ϵ a zero-mean Gaussian distributed random variable with standard deviation σ_ϵ (in dB) that represents the shadowing effects.

The received signal strength (P_r) at a distance d is the output power of the transmitter minus $PL(d)$. Formally:

$$P_r(d) = P_t - PL(d) = P_t - PL(d_0) - 10\gamma \log_{10}(\frac{d}{d_0}) - \epsilon \qquad [1.8]$$

Figure 1.4 shows an analytical propagation model for $\gamma = 2$, $\sigma_\epsilon = 4$, $PL(d_0) = 55$ dB, $d_0 = 1$, and an output power $P_t = 0$ dBm (e.g. the Chipcon CC2420 IEEE 802.15.4, 2.4 GHz).

From equation [1.8] we have $P_r(d) \sim \mathcal{N}(P_t - PL(d_0) - 10\gamma \log_{10}(\frac{d}{d_0}), \sigma_\epsilon)$. Since $P_r(d)$ is Gaussian, the probability of successful communication between two sensors s_i and s_j located at distance d from each other is:

$$\mathcal{P}[P_r(d) > SS_{min}] = Q(\frac{SS_{min} - (P_t - PL(d_0) - 10\gamma \log_{10}(\frac{d}{d_0}))}{\sigma_\epsilon}) \qquad [1.9]$$

where SS_{min} represents the minimum acceptable signal strength and Q is the complementary cumulative distribution function (CCDF) of a standard Gaussian, i.e.

$$Q(x) = \frac{1}{\sqrt{2\pi}} \int_x^{+\infty} e^{\frac{-t^2}{2}} dt$$

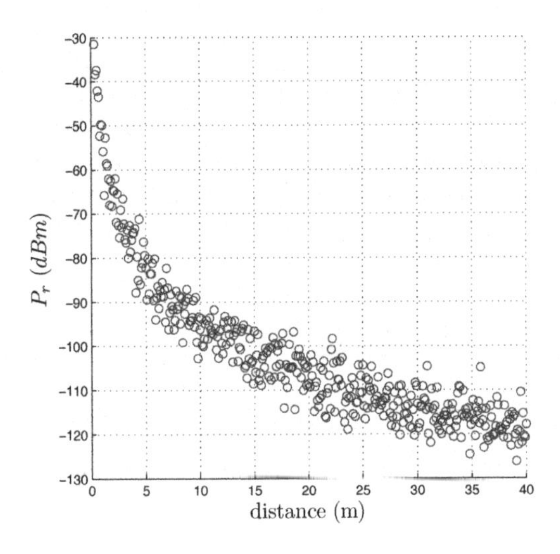

Figure 1.4. *Channel model,* $\gamma = 2$, $\sigma_\epsilon = 4$, $P_t = 0$ *dBm*

Figure 1.5 shows the connectivity model related to this formulation. We see clearly that some areas within the connectivity range receive power lower than SS_{min}. On the other hand, some areas outside the connectivity range receive power higher than SS_{min}. As mentioned before, this model reflects the fact that there is no clear cut-off boundary between successful and unsuccessful communication.

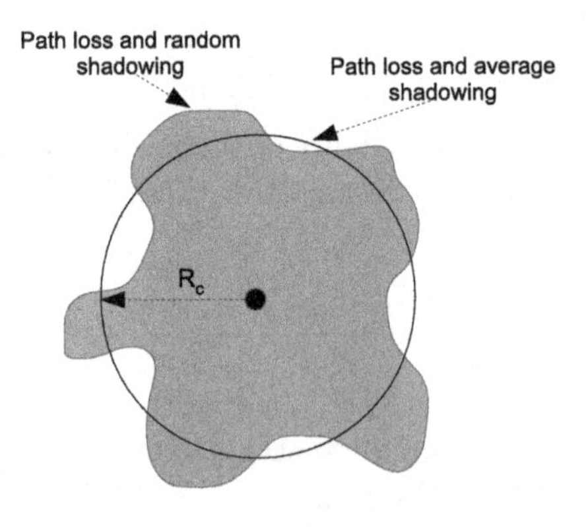

Figure 1.5. *Connectivity model*

1.1.4. *Wireless sensor networks*

A WSN consists of spatially distributed sensors, and one or more sink nodes (also called base stations). Sensors monitor, in real-time, physical conditions, such as temperature, vibration, or motion, and produce sensory data. A sensor node could behave both as data originator and data router. A sink, on the other hand, collects data from sensors. For example, in an event monitoring application, sensors are required to send data to the sink(s) when they detect the occurrence of events of interest. The sink may communicate with the end-user via direct connections, the Internet, satellite, or any type of wireless links. Figure 1.6 depicts a typical WSN architecture. Note that there may be multiple sinks and multiple end-users.

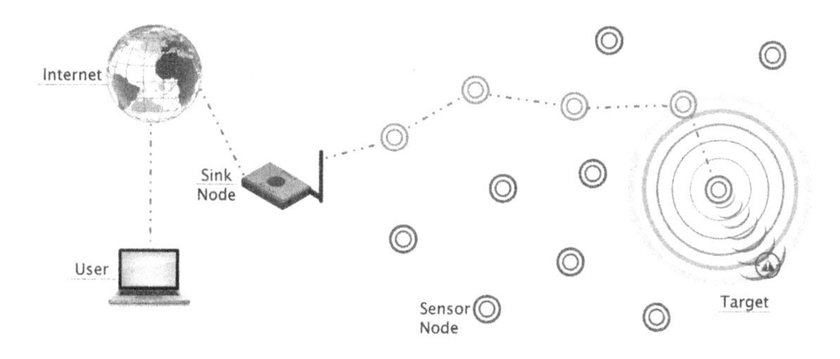

Figure 1.6. *Typical WSN architecture. For a color version of the figure, see www.iste.co.uk/senouci/wireless.zip*

As a fundamental issue in WSNs, deployment is a research topic that has attracted much attention in recent years [DHI 02, ZOU 03b, VIE 04, LIN 05, ZHA 06, WU 07a, YOU 08, AIT 09, WAN 10a, ABA 11, BHU 12, AKB 13, AMM 14, SEN 15c]. Indeed, the number and locations of sensors, deployed in a RoI, determine the topology of the network, which will further influence many of its intrinsic properties, such as its coverage, connectivity, cost and lifetime. Consequently, the performance of a WSN depends to a large extent on its deployment.

In practice, special cases of WSNs are encountered such as wireless multimedia sensor networks (WMSNs), underwater wireless sensor networks (UWSNs), wireless underground sensor networks (WUSNs), wireless body sensor networks (WBSNs) and wireless sensor-actor networks (WSANs).

1.1.5. *WSN models and architectures*

Depending on the sensors' capabilities and application requirements, sensors can cooperate according to various models and architectures. In what follows, we introduce some typical WSN models and architectures:

– *small-, medium-, large- and very large-scale WSNs*: the size of the WSN varies depending on several factors such as the sensors' characteristics, the RoI, and the user's requirements. In practice, the number of sensor nodes in a WSN may be in the order of tens, hundreds, thousands, or even tens of thousands;

– *homogeneous versus heterogeneous WSNs*: a WSN may be homogeneous or heterogeneous. A WSN is homogeneous if all sensors of the network have the same capabilities (sensing, processing, communication, etc). A heterogeneous WSN consists of sensors endowed with different capacities, which may serve for different applications. Typically, some sensors will have more resources available, such as processing and energy, than the rest of the sensors;

– *stationary, mobile, and hybrid WSNs*: a WSN may be stationary, mobile, or hybrid. A stationary WSN is a network consisting of stationary sensor nodes that cannot move once deployed. With the advances in mobile devices, some of the sensors are able to move on their own; this is generally achieved by embedding the sensors on mobile platforms (Figure 1.7). A mobile WSN comprises only mobile sensors, while a hybrid WSN consists of both stationary and mobile sensors;

Figure 1.7. *Photograph of an Arduino-based mobile sensor node*

– *flat versus hierarchical WSNs*: in flat WSNs, all the sensor nodes are assumed to be homogeneous and play the same role. However, in hierarchical WSNs, a sensor node can be dedicated to a particular special function. For instance, a sensor could be designated as a cluster-head, in charge of communicating with adjacent clusters;

– *single-hop versus multi-hop WSNs*: in a single-hop WSN, sensor nodes transmit their data directly to the sink. In a multi-hop WSN, multiple relaying sensor nodes exist between sensors and sinks. A multi-hop WSN can be flat or hierarchical.

1.1.6. *WSN applications*

A WSN may include different types of sensors to monitor almost any ambient condition, which open the doors for a wide range of applications in different domains, such as: military, industrial, environmental, home and medical applications. Typical applications of WSNs include, but are not limited to:

– *military applications*: the ability to deploy unmanned surveillance missions, by using WSNs, is of great practical importance for the military. WSNs have been identified as an ideal alternative to conventional surveillance systems that provide an essential component of battlefield awareness. Indeed, in recent years, WSNs have been applied in various military applications, such as monitoring friendly forces, shooter localization, battlefield surveillance, and battle damage assessment [ARO 04, LED 05, HE 06, VIC 09, GEO 13]. For instance, VigilNet [HE 06] is a military WSN that acquires and verifies information about enemy capabilities and positions of hostile targets. It has been successfully designed, built, demonstrated and delivered to the Defense Intelligence Agency for realistic deployment;

– *industrial applications*: recent advancements in sensor technology have made WSNs prevalent in numerous industrial applications such as structural health monitoring (e.g. the Guangzhou New TV Tower, China, and the Ting Kau Bridge, Hong Kong [NI 08]), pipeline monitoring (e.g. TriopusNet [LAI 12]), thermal monitoring in data centers [CHE 14c], oil refineries monitoring (e.g. GINSENG [O'DO 13]), and agriculture crop monitoring [JUU 15]. For instance, in [JUU 15] a WSN was deployed to automate the process of monitoring crop storages and ensure proper storage conditions. Figure 1.8 shows a photograph of employed sensors;

Figure 1.8. *The node and the protective shell [JUU 15]*

– *environmental applications*: WSNs have widely been used for environmental and wildlife monitoring [MO 09, DYO 10]. Some environmental applications of WSNs include tracking the movements of birds or animals, automatic irrigation, precision agriculture, and pollution studies [ZHA 04b, WER 06, KIM 08]. As an example, WSNs are employed in FIRESENSE [FIR 15], which is a Specific Targeted Research Project (STReP) of the European Union's 7th Framework Programme Environment. FIRESENSE aims to develop an automatic early warning system to monitor areas of archaeological and cultural interest remotely from the risk of fire and extreme weather conditions;

– *home applications*: WSNs can help to create a smart environment by interconnecting various devices in residential spaces with convenient control of various applications at home. An example of smart environments is the smart kindergarten project [CHE 02], which uses WSNs to create a smart environment for early childhood education;

– *healthcare applications*: WSNs have enabled a rapid development of telemedicine systems, which provide remote monitoring of patients and their vital parameters [YAN 06a]. Some potential medical applications include real-time continuous patient monitoring, home monitoring for chronic patients, and collection of long-term databases of clinical data. An example project is the CodeBlue [LOR 04] that develops a WSN for monitoring and tracking of patients. Also, infirm and elderly people can benefit greatly from healthcare applications of WSNs, bringing a better quality of life and even saving lives through WSN technology. For instance, AlarmNet [WOO 08] is a WSN designed for long-term health monitoring in assisted-living environments. AlarmNet is extensible and adapts to the individual context and behavior patterns of the residents.

1.2. WSN deployment strategies

One of the fundamental design issues in WSNs is where to place the sensors in the RoI. The location of a sensor may affect the fulfillment of the system's requirements and multiple network performance metrics. Careful sensor placement can be a very effective optimization means for achieving the desired design goals. For example, the coverage objective regards how to ensure that each of the target points in the RoI is covered by the WSN. The sensors need to be placed neither too close to each other so that the sensing capability of the network is fully utilized and at the same time not too far from each other to avoid the formation of coverage holes. A good deployment will enable a better performance on information gathering and communication.

Sensors can generally be placed in the RoI either deterministically or randomly. The choice of the deployment scheme depends highly on the type of sensors, application, and the environment that the sensors will operate in [YOU 08, WAN 09, WAN 10a]. Deterministic deployment schemes are optimal as sensors are placed at predetermined coordinates to guarantee network efficiency; however, they are impractical and sometimes impossible for large-scale WSN applications. For these applications, sensors may be deployed from a plane, delivered in an artillery shell, rocket or missile, or catapulted from a shipboard [AKY 02]. In these cases, the WSN has the utmost challenge of guaranteeing proper area coverage and connectivity upon deployment [AKY 02, YOU 08]. Such cases require implementation of additional complex protocols to ensure efficient network operation, which maximize network lifetime and decrease frequency of re-deployment. Random deployment of WSNs will be discussed in Chapter 2. Chapters 3 and 4 discuss deterministic deployment and fusion-based deterministic deployment, respectively.

Using mobile sensors, some deployment strategies have advocated dynamic adjustment of sensors' location which allows a dynamic deployment reconfiguration that improves network performance. One of the major techniques is to schedule mobile sensors to move to the designated locations according to the results computed by the placement strategy such that the energy consumption due to movement is minimized. Another major approach is to schedule mobile sensors in order to cope with coverage hole issues. Chapter 5 discusses dynamic, deployment.

All the WSN deployment strategies mentioned above should take into account the optimization of one or more objectives relating to the application needs under one or more constraints. In the following, we will discuss such objectives and constraints.

1.3. WSN deployment: objectives and constraints

This section introduces the objectives and constraints generally considered in the literature.

1.3.1. *Coverage*

The first challenge encountered in WSNs is how to cover a RoI perfectly. Coverage is one of the most fundamental issues in WSNs, which have a great impact on the performance of WSNs [WAN 11]. Generally speaking, coverage is a measure of the quality of service (QoS) of the sensing function and is subject to a wide range of interpretations due to a large variety of sensors and applications. Several coverage formulations arise naturally in many domains. Typically considered formulations are point coverage, target coverage, barrier coverage, and area/blanket coverage. The point coverage approaches seek surveillance solutions that find stationary objects within the RoI, whereas the target coverage approaches look to find targets as they move through the RoI. This type of coverage has obvious military applications such as those covered in [ARO 04]. A variation of point/target coverage known as sweep coverage is also discussed in [LI 11]. Barrier coverage refers to the detection of movement across a barrier of sensors. This problem was defined as the maximal breach path in [MEG 01]. In this book, we focus on area coverage, by which we are interested in the coverage of the whole RoI.

Assessing the coverage varies based on the underlying sensor coverage model and the metric system used to measure the collective coverage of deployed sensors. Some of the published papers use the worst case coverage, usually referred to as least exposure, measuring the probability that a target would travel across the RoI or an event would happen without being detected. However, most of the works have focused on the ratio of the covered area to the size of the RoI as a metric for the rate of coverage.

In the case of detection applications, the performance of a WSN is usually specified in terms of its overall detection and false alarm rate. Detection rate refers to the probability that a "yes" detection decision is made when the phenomena of interest is present. Obviously, correct detection decisions are always desirable. False alarm rate refers to the probability that a detection decision of "yes" is made when the phenomena of interest is absent. False alarms are undesirable but can occur due to imperfections in sensor design or noise. Maximizing the overall detection rate and minimizing the overall false alarm rate are competing objectives. Therefore, the end user of a WSN detection system usually provides a set of minimum detection requirements to satisfy and a set of false alarm requirements that are not to be exceeded.

It should be noted that in the literature there is confusion concerning the difference between coverage, or coverage rate, and the quality of coverage. Existing literature does not always make a clear distinction between the two terms which are often used interchangeably. In this book, a distinction is made between these two terms. Coverage or coverage rate refers to the ratio of the covered area, in terms of detection rate to the size of the RoI, whereas the quality of coverage takes into account the false alarm rate in addition to the coverage rate.

1.3.2. *k-coverage*

Applications requiring k-coverage ($k > 1$) may occur in situations where a stronger environmental monitoring capability is desired, such as military applications. Such a problem can be formulated as a decision problem whose goal is to determine whether every point in the RoI is covered by at least k sensors, where k is a predefined value.

1.3.3. *Preferential coverage*

The preferential/differentiated/non-uniform coverage, which considers the satisfaction of different coverage levels in different geographical areas of the

RoI, is also an important issue. The distribution of the requested event detection probability will not be uniform as it will depend on the application, the RoI and the event characteristics. The requested event-detection probability distribution can also be influenced by geographical factors. In many real world WSN applications, such as surveillance applications, the supervised RoI may require high detection probabilities in certain sensitive areas. However, for some not so sensitive areas, relatively low detection probabilities are required to reduce the number of deployed sensors (e.g. Figure 1.9). For example, in a fire-detection system, high detection probabilities (close to 1) are requested for high-risk areas (e.g. those close to chemical deposits), and low detection probabilities for low-risk areas. In this case, the coverage rate is the percentage of the RoI where the generated detection probabilities are equal or greater than the required detection probabilities.

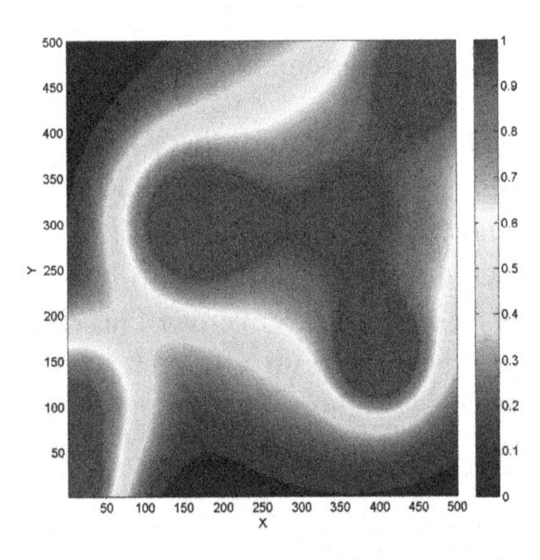

Figure 1.9. *Non-uniform coverage scenario. For a color version of the figure, see www.iste.co.uk/senouci/wireless.zip*

1.3.4. *Deployment cost*

The more sensors are used the higher is the overall cost of the network. As the latter must be taken into consideration, the number of deployed sensors is one of the important metrics that needs to be considered within WSN deployment processes. Almost all the papers dedicated to optimal sensor deployment consider achieving the specified goals with minimum cost. In homogeneous WSNs, the deployment cost is the number of deployed sensors, hence reducing the deployment cost is achieved by reducing the number of deployed sensors while satisfying the user requirements. However, in heterogeneous WSN, deployment cost is defined as the sum of costs of the different deployed sensors. As different sensors have different costs, reducing the deployment cost is achieved by finding the best combination that ensures the application needs with the minimum cost.

1.3.5. *Network connectivity*

Another issue in WSN deployment is the connectivity of the network. Usually, a network is connected if any sensor can communicate with any other sensor (possibly using other sensors as relays: multi-hop path). This definition can be relaxed with regards to communication among sensors: one-to-one, one-to-many, and many-to-one. For instance, in surveillance applications, only many-to-one communication is required. Network connectivity is necessary to ensure that the information collected by sensors is sent to the sink. In this case, we say that a network is connected if any sensor can communicate with the sink (either directly or via a multi-hop path).

It should be noted that connectivity and coverage are related since they are both affected by the position of sensors. This relationship will be discussed in subsequent chapters.

1.3.6. *Network lifetime*

One major challenge in the design of WSNs is the fact that energy resources are very limited. Recharging or replacing the battery of deployed sensors may be difficult or impossible. The number and positions of sensors have a direct impact on the WSN lifetime. When few sensors are deployed, a weak lifetime results. However, deploying many sensors does not automatically increase the lifetime. In fact, when the number of sensors increases, volume of traffic also increases. Consequently, the energy consumption increases and the lifetime decreases. Thus, deploying more sensors does not necessarily solve the problem. To increase the network lifetime, sensors must be located in the right places by taking into consideration factors such as the position of the sink(s), event frequency, and any specific mechanism implemented in the sensors to build paths to the sink (routing protocol).

1.3.7. *Energy efficiency*

This criterion is often used interchangeably with the network lifetime. Due to the limited energy resource in each sensor, we need to use the sensors in an efficient manner to increase the network lifetime. Energy efficiency is considered as the major challenge for the proliferation of WSNs [AKY 10]. There are at least two approaches to the problem of conserving energy in sensor networks connected with optimal placement [KHA 15]. The first approach is to plan a schedule of active sensors that enables other sensors to go into sleep mode using overlaps among sensing ranges. The second approach is adjusting the sensing range of sensors for energy conservation.

In the case of mobile sensors, energy efficiency mainly consists of scheduling mobile sensors to move to the designated locations according to the results computed by the placement strategy such that energy consumption due to movement is minimized.

1.3.8. *Data fidelity*

Ensuring the credibility of the gathered data is obviously an important design goal of WSNs. A sensor network basically provides a collective assessment of the detected phenomena by fusing the readings of multiple independent (and sometimes heterogeneous) sensors. Data fusion boosts the fidelity of the reported incidents by lowering the probability of false alarms, and of missing a detectable object. Increasing the number of sensors reporting in a particular region will surely boost the accuracy of the fused data. However, redundancy in coverage would require an increased sensor density, which can be undesirable due to increased cost or decreased survivability (e.g. the potential of detecting the sensors in a combat field).

1.3.9. *Fault tolerance and load balancing*

Since WSNs are subject to failures, fault-tolerance becomes an important requirement for many WSN applications [CHO 15]. In fact, the availability of the services provided by a WSN is to a large extent affected by faults that may occur due to various reasons, such as malfunctioning, hardware and software glitches, dislocation, or environmental hazards, e.g. fire or flood. A WSN that is not prepared to deal with such situations may suffer a reduction in overall lifetime, or lead to hazardous consequences in critical application contexts.

One of the major fault tolerance techniques is the exploitation of redundancy, which is often a default condition in WSNs. To deal with failures, techniques for reliable routing in WSNs have been proposed and are well-understood, but to be effective, these techniques depend on the physical network topology that should ensure that alternative routes to the sink are available. This requires WSN deployment to be planned with an objective of ensuring some measure of robustness in the topology, so when failures occur the routing protocols can continue to offer a reliable delivery. Thus, a WSN should be k-connected, which allows $k - 1$ sensors to fail while the WSN would still be connected. A second approach is sensor value fusion that seeks to provide high-level information derived from a number of low-level sensor inputs. There, the inherent redundancy of sensors that can be used to provide fault-tolerant data aggregation. Another major approach is the involvement of sinks or other resourceful sensors to maintain operations after failures.

1.4. Conclusion

In this chapter, sensors, WSNs, deployment strategies and the objectives and constraints related to the deployment of WSNs are discussed. Analytical sensor coverage and sensor communication models are also described to model the detection and communication in WSNs. In the next chapter, we will discuss the most *naive* approach to deploy WSNs: random deployment.

2

Random Deployment

This chapter focuses on random deployment of WSNs, which is often employed in remote, or hostile environments. To provide a holistic view of random deployment, this chapter presents a survey and taxonomy of random deployment strategies. Moreover, it discusses published theoretical and simulation studies on random deployment. Finally, it formulates practical issues that need further research.

2.1. Why random deployment?

In the literature, sensor deployment is typically carried out in either of the two approaches: *deterministic deployment* or *random deployment*. The selection of a suitable approach depends on many factors, such as the type of sensors, the nature of the RoI, and the application needs [AKY 02, YOU 08]. Deterministic deployment schemes are optimal as sensors are placed at predetermined coordinates to guarantee network efficiency. However, they are impractical and sometimes impossible due to the large scale of the WSN, or to the inaccessibility/harshness of the RoI. For instance, in harsh environments such as a battlefield, or a disaster region, deterministic deployment of sensors is very risky and/or infeasible. In this case, random deployment often becomes the only option, sensors may be deployed from a plane, delivered in an artillery shell, rocket or missile, or catapulted from a shipboard [AKY 02].

When there is a high density of sensors, various adjusting mechanisms can be adopted to turn off redundant sensors in order to improve the WSN lifespan. Activity scheduling mechanisms [XIN 05, YAN 08a] control the active and sleep states of the sensors, a subset of the deployed sensors can be chosen dynamically to remain active to maintain certain network properties such as coverage and connectivity, while ensuring longer network lifetime through energy conservation and balancing among sensors.

Random deployment is also an essential component in other kinds of deployment. For instance, dynamic deployment of WSNs (see Chapter 5) that deals with mobile sensors, often assumes that initially sensors are randomly deployed [SEN 15b]. Afterward, it strives to efficiently move the mobile sensors so that the final deployment meets the design goals of the network.

2.2. Random deployment strategies

Random deployment is the most practical way of placing sensors. In random (or stochastic) node placement, sensor-positions are defined by a probability density function (PDF)[1]. Depending on the deployment strategy, the coordinates of the sensor positions may follow a particular distribution. In this section, we define the PDF and briefly describe the characteristics of random node placement strategies found in the literature. We categorize the random placement strategies into *simple* and *compound* (Figure 2.1). Simple strategies are mere variants of the simple diffusion strategy, whereas compound strategies are realized by repeated simple diffusion.

2.2.1. *Simple random node placement strategies*

2.2.1.1. *Simple diffusion*

The simplest way to deploy sensors is to scatter them from the air [AKY 02, ISH 04b, WAN 08a]. Since all the information must reach the sink, the distribution is centered on the sink. Lightweight sensors will have higher air resistance randomizing their placement and the resulting distribution is called *simple diffusion*. In [ISH 04a], this deployment process was modeled by a linear diffusion equation, whose solution is a two-dimensional normal

1 The probability of a sensor being within the RoI = $\{x_1 \leq X \leq x_2, y_1 \leq Y \leq y_2\}$ can be written in terms of PDF as follows: $P(x_1 \leq X \leq x_2, y_1 \leq Y \leq y_2) = \int_{x_1}^{x_2} \int_{y_1}^{y_2} f(x,y)dxdy$.

distribution. The PDF of sensor-positions is:

$$f(x) = \frac{1}{2\pi\sigma^2} h(||x - c||), \quad x \in \mathbb{R}^2 \qquad [2.1]$$

$$h(r) \triangleq e^{\left(\frac{-r^2}{2\sigma^2}\right)}$$

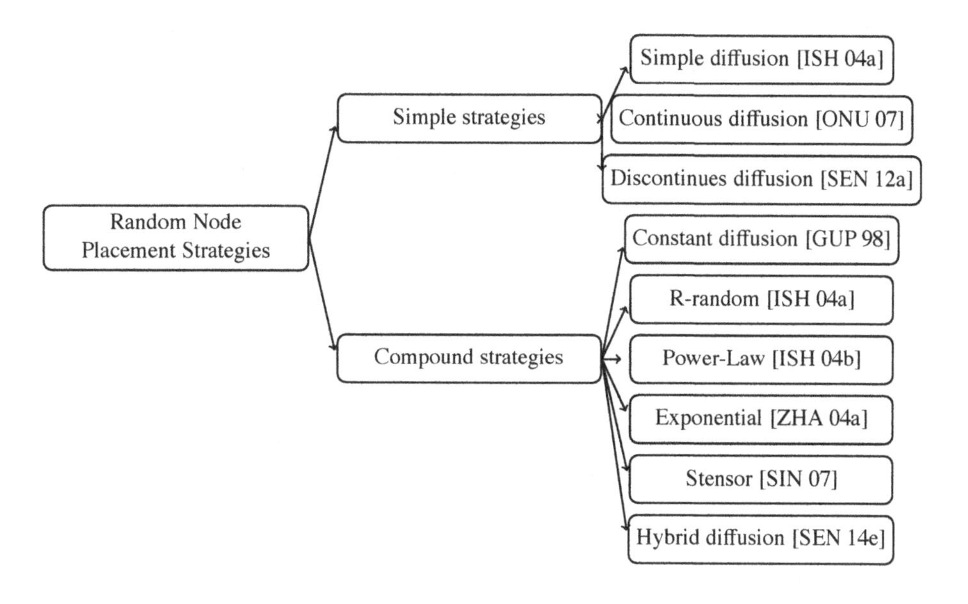

Figure 2.1. *Random node placement strategies taxonomy*

In equation [2.1], $c \in \mathbb{R}^2$ is the mean position at ground, which is just under the point where sensors are scattered, and σ^2 is the variance of the distribution. The variance is determined by various factors (e.g. shape or weight of sensors, or the height from which sensors are released). It should be pointed out that when air currents are strong, sensor-positions are governed by another formula [ISH 04a]. Figure 2.2 shows an example of a simple diffusion of 498 nodes in a RoI of 300 m × 300 m.

Simple diffusion is a very important random deployment strategy. In fact, it has been shown that any random node placement can be realized by repeated simple diffusions with different means and variances. The theoretical basis for this consideration is described in [ISH 04b].

Figure 2.2. *Example of simple diffusion*

2.2.1.2. *Continuous diffusion*

If sensors are thrown off an aircraft that flies over the middle of the RoI, most sensors are expected to fall somewhere close to the central line (denoted by B), and several sensors are likely to end up further out. Along the axis of flight, the node distribution is uniform, while it is Gaussian in the orthogonal direction [ONU 07]. The resulting distribution is called *continuous diffusion*. The PDF of sensor-positions is (the x-axis is the axis of flight):

$$f(x) = \frac{1}{|B|}, \quad x \in \mathbb{R} \qquad [2.2]$$

$$f(y) = \frac{1}{\sqrt{2\pi\sigma^2}} e^{-\frac{(y-m)^2}{2\sigma^2}}, \quad y \in \mathbb{R}$$

where $|B|$ is the length of the axis of flight. Figure 2.3 shows an example of a continuous diffusion of 497 sensors in a RoI of 600 m × 200 m.

2.2.1.3. *Discontinuous diffusion*

In this model, the sensors are dropped by an aircraft that flies over the middle of the RoI. Senouci *et al.* [SEN 12a] propose a discontinuous dropping of the sensors defined as follows: sensors are thrown discontinuously in a single flying-over. In each throw, n sensors will be dropped. In the end, we will have N dropped sensors, with

$N = n \times$ *number of throws*. Figure 2.4 shows an example of a discontinuous diffusion in a RoI of 600 m × 200 m, with $N = 98$ and the *number of throws* = 5. If the number of throws is increased, the discontinuous diffusion converges to the continuous diffusion, as shown in Figure 2.5.

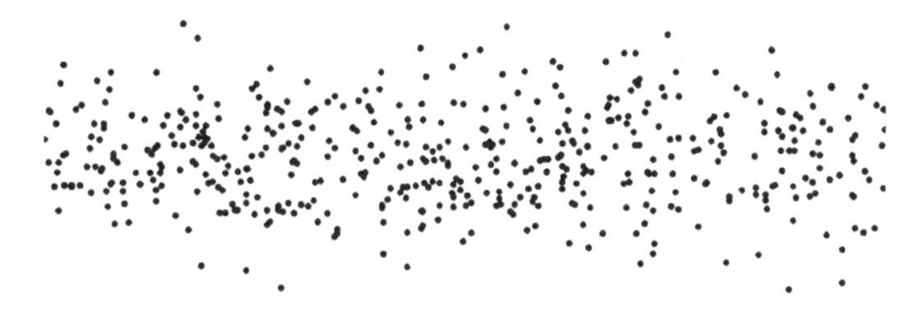

Figure 2.3. *Example of continuous diffusion*

Figure 2.4. *Example of discontinuous diffusion (5 throws)*

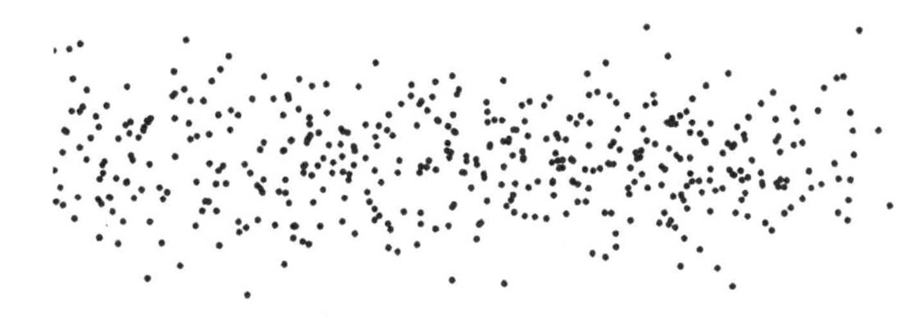

Figure 2.5. *Example of discontinuous diffusion (10 throws)*

2.2.2. Compound random node placement strategies

A compound random deployment model is a random deployment model realizable by repeated simple diffusions with different means and variances.

Therefore, to practically obtain a compound random deployment model we may need to fly over the RoI several times. In this section, we survey some existing compound random node placement strategies.

2.2.2.1. *Constant diffusion*

In many works [GUP 98, YOU 08], the sensors are placed in such a way that their density is constant. Such random distribution is called *constant diffusion*. The PDF of the sensor-positions is given by the following equation:

$$f(x) = \frac{1}{|RoI|}, \quad x \in \mathbb{R}^2 \qquad [2.3]$$

An example of constant diffusion is illustrated in Figure 2.6. Here, the number of sensors is 400 and the RoI is 300 m × 300 m.

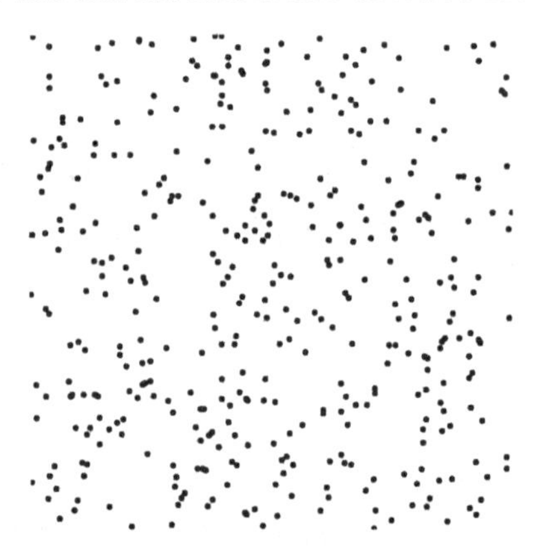

Figure 2.6. *Example of constant diffusion*

2.2.2.2. *R-random*

This distribution was proposed in [ISH 04a], where the nodes are uniformly scattered with respect to the radial and angular directions from the sink. The R-random node distribution pattern simulates the effect of an exploded shell and follows the following PDF for sensor-positions in polar-coordinates within a distance R from the sink:

$$f(r, \theta) = \frac{1}{2\pi R} \quad 0 \leq r \leq R, \ 0 \leq \theta \leq 2\pi \qquad [2.4]$$

An example of R-random placement is illustrated in Figure 2.7. Here, the number of sensors is 394 and the RoI is 300 m × 300 m.

Figure 2.7. *Example of R-random*

2.2.2.3. *Power-law*

This distribution is characterized by the following two features [ISH 04b]: first, the density of sensors is higher near the sink, and second, the degree of the sensors follows a power law. The PDF of the sensor-positions in polar-coordinates is:

$$f(r,\theta) = \frac{\alpha+1}{2\pi R}(\frac{r}{R})^{\alpha} \qquad [2.5]$$

$$0 \le r \le R,\ \ 0 \le \theta \le 2\pi,\ \ -1 \le \alpha \le 1$$

The characteristics of the Power-law placement are similar to those of the R-random placement.

2.2.2.4. *Exponential*

In this model, the distribution follows an exponential law [ZHA 04a]. The PDF of sensor-positions is:

$$f(x) = \lambda e^{-\lambda x} \qquad [2.6]$$

An example of exponential diffusion is illustrated in Figure 2.8. Here, the number of sensors is 436, the RoI is 300 m × 300 m, and $\lambda = 100$.

Figure 2.8. *Example of exponential diffusion*

2.2.2.5. *Stensor*

In [SIN 07], the authors propose Stensor, a partition-based random node placement algorithm. They assumed a rectangular area classified into small cells. Each cell cannot host more than one sensor, and sensors are distributed in those cells according to the following PDF:

$$f(x) = \frac{e^{-\sqrt{\lambda}} \lambda^{\frac{x}{2}}}{x!}, \quad x \geq 0 \qquad [2.7]$$

An example of Stensor placement is illustrated in Figure 2.9. Here, the number of sensors is 16 and the RoI contains 1000 cells.

Figure 2.9. *Example of Stensor placement [SIN 07]*

2.2.2.6. *Hybrid diffusion*

Senouci *et al.* [SEN 14e] proposed a hybridization of the simple diffusion model that places a large number of nodes around the sink and the constant diffusion that provides high coverage and connectivity rates. They chose simple diffusion instead of the R-random as the latter ensures a greater detection rate, because the simple diffusion is characterized by the standard deviation σ which makes it possible to control the density of nodes around the sink. The authors call such hybridization the *hybrid diffusion*.

The hybrid diffusion of N nodes is defined as the deployment of $\alpha.N$ nodes according to the simple diffusion strategy and $\beta.N$ nodes according to the constant diffusion model where $0 < \alpha,\beta < 1$ and $\alpha + \beta = 1$. Figure 2.10 shows an example of a hybrid diffusion of 300 nodes in a circular RoI of a radius $R = 150$ m.

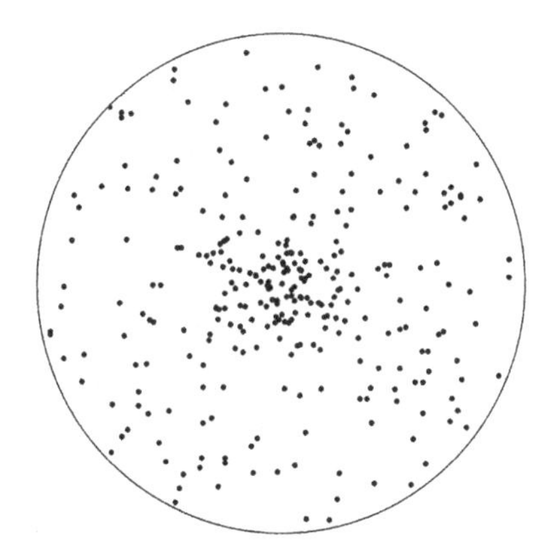

Figure 2.10. *Example of hybrid diffusion*

2.3. Discussion

Due to its practicability, random deployment has been studied extensively over the last decade. Generally speaking, studies on random deployment can be classified into two types: theoretical studies and empirical studies. With both approaches, the effectiveness of random deployment has been shown to be certain. In this section, we provide a discussion on the different published works related to the random deployment of WSNs.

2.3.1. *Theoretical studies*

Due to the analytical complexity of random deployment strategies, theoretical studies are usually conducted for very few random deployment models under several simplifying assumptions [GUP 98, ZHA 04a, WAN 06a, WAN 08a, AMM 08, VAL 13, SEV 14]. In these studies, the WSN is often modeled by means of random geometric graphs (RGGs) [PEN 03], and the number of nodes is assumed infinite (large-scale WSNs). In addition, the percolation theory [AMM 08, SEV 14] is largely adopted to analyze coverage and/or connectivity. Asymptotic results reported by theoretical studies are very important, as they show some fundamental trade-offs in WSNs, and also give us good estimate for large-scale WSNs.

Gupta and Kumar [GUP 98] studied necessary conditions on the transmission range needed for asymptotic connectivity of constant diffusion. Under the assumptions that nodes are deployed as a Poisson point process in a square region, the authors in [ZHA 04a] derived the node density required in order to maintain full coverage with high probability. Assuming the same distribution, Ammari and Das [AMM 08] analyzed the relationship between coverage and connectivity. Wan *et al.* [WAN 06a] studied how the probability of the k-coverage changes with the sensing radius or the number of sensors. They assumed that the sensors are deployed as either a Poisson point process or a uniform point process in a square region. Wang *et al.* [WAN 08a] identified intrinsic properties of coverage and lifespan of a WSN that follows the two-dimensional Gaussian distribution. They showed that Gaussian distributions can effectively increase the network lifespan. Recently, Eslami *et al.* [ESL 13] showed that the asymptotic results reported by the abovementioned studies provide poor approximations for finite (small-scale) WSNs. The authors considered randomly deployed finite WSNs, and derived lower and upper bounds for their coverage and k-connectivity.

Some recent studies consider clustered WSNs. Authors in [VAL 13] studied the case where several clusters of sensors are spread over the region of interest following Gaussian random distributions. Sevgi and Koçyigit [SEV 14] analyzed partial connected coverage of statically clustered and heterogeneous WSNs.

2.3.2. *Simulation studies*

In contrast to theoretical studies, simulation studies consider a set of metrics and several random deployment models [ISH 04a, ISH 04b, ONU 07, SEN 12a, SEN 14e]. In [ISH 04a], the authors evaluated the fault-tolerance against random failure from the random deployment viewpoint. Results prove that the tolerance against failure is low in constant placement when the R-random placement has high fault-tolerance. In another paper [ISH 04b], the same authors showed that the Power-law placement can raise fault-tolerance with appropriately selected control parameters.

In [VAS 09], the authors discussed a performance study of congestion control algorithms when nodes are deployed under four different topologies, namely: simple diffusion, constant placement, R-random placement and grid placement. The results depict that congestion control algorithms are highly affected by the node placement. In a more recent paper [SER 10], the same authors evaluated the energy utilization performance of a congestion control and avoidance algorithm while considering the grid placement, biased random placement, simple diffusion and random placement. Obtained results show that the best performance is obtained when the sensors are densely deployed near hot spots like the sink.

Senouci *et al.* [SEN 14e] conducted a simulation study to analyze the intrinsic properties of several random node placement strategies such as: constant diffusion, continuous diffusion, R-random diffusion, simple diffusion exponential diffusion and hybrid diffusion. They consider the following metrics: coverage, connectivity, connected coverage, network lifespan, fault-tolerance and routing-related metrics. The obtained results give helpful design guidelines in using random deployment strategies.

2.4. Practical issues that need further research

Although, random deployment is a classic problem that was studied long ago for air warfare [LAU 57], it is still a hot research topic. Significant progress has been made, but many challenging problems remain. In this section, we highlight open research problems, identify the issues involved, and report on ongoing work and preliminary results.

2.4.1. *Finite WSNs*

Many analytic asymptotic results for coverage and connectivity of WSNs have been reported for the case of infinite WSNs (large-scale WSNs). However, in many real-life WSNs, the size of the WSN may be limited to a few hundred sensor nodes (small-scale or finite WSNs). Evidence reported in [ESL 13] shows that the asymptotic results are not suitable for analyzing finite WSNs. Thus, future research in this area should consider extending the asymptotic results for WSNs with practical sizes.

2.4.2. *Realistic sensor coverage and communication models*

An important yet underrated issue in WSNs is that random deployment not only randomizes sensors positions but also sensors postures, which may lead to adverse positions (e.g. upside-down and side-facing). The effect of such positions on the WSN performance could be dramatic. Indeed, adverse positions generate random antenna orientations leading to network performance degradation [WAD 09]. This is also true for many sensing circuitry. Therefore, this issue should be considered carefully within the WSNs' random deployment process.

A straightforward approach to tackle the issue discussed above is to consider more realistic sensor coverage and communication models. In fact, there is an unquestionable need for coverage and radio models that are close enough to reality and effectively characterize the effect of random deployment, while at the same time being concise enough to promote strong theoretical results. Indeed, some preliminary research results have started to emerge. In a recent work, Won *et al.* [WON 14] presented a radio model that captures the characteristics of random deployment. This model was built upon the basis of the radio irregularity model (RIM) [ZHO 06].

2.4.3. *Practical sensor package for random deployment*

As discussed previously, very important results regarding random deployment were reported in the literature using both theoretical and simulation studies. However, investigations on real-world random deployment related-issues seem to be relatively slow. For instance, the research community has totally ignored issues related to sensor packaging supposed to be used in real-life deployment. Although, many sensor packages were designed to effectively protect sensors from the physical damages during random deployment, they simply ignore the issue of heterogeneous sensor postures.

Recently, Won *et al.* [WON 14] presented a simple sensor package that ensures the upright position of a sensor. The idea is to place a heavy material at the bottom part of the sensor package to prevent it from overturning (Figure 2.11). Various other approaches in designing sensor packages are expected in the near future.

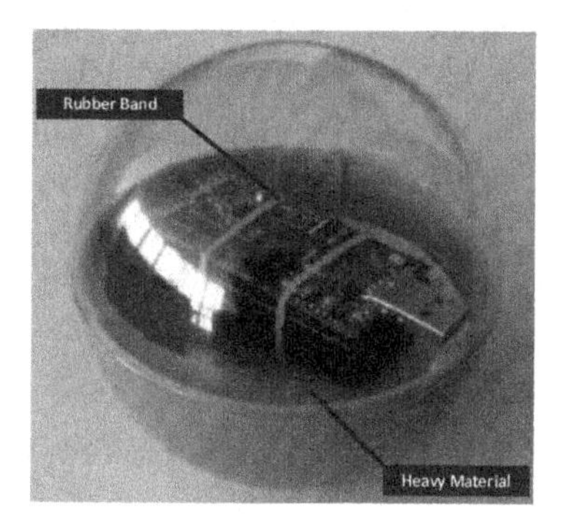

Figure 2.11. *Prototype of the sensor package RolyPoly [WON 14]*

2.4.4. *Random deployment of three-dimensional WSNs*

The increasing interest in using WSNs for underwater surveillance applications underscores the importance of investigating random deployment of WSNs in three-dimensional spaces. In fact, in underwater environments, deploying and maintaining sensors at pre-computed locations is very difficult, if not impossible. Thus, in such environments, a reasonable assumption is to assume that the positions of sensors are random and uncontrollable.

Alam and Haas [ALA 15] investigated the coverage and connectivity issues in three-dimensional WSNs in situations where it was difficult to deploy and maintain nodes in predetermined positions. The authors partition the 3D network space into identical regions, and keep one node active in each such cell. Among different partitioning schemes, the authors found that the truncated octahedron-based partitioning scheme is the best choice.

2.5. Conclusion

As discussed earlier, the deployment is a mandatory and critical step in the process of developing WSN solutions for real-life applications. In this chapter, we have discussed random deployment, which is often the best choice when considering inaccessible or harsh RoI, and/or large-scale WSNs. Although random deployment is widely used in theoretical and simulation studies, it has often been considered too expensive in comparison to the optimal deterministic deployment. In fact, it is known that deterministic deployment will need $O(\log n)$ times fewer sensors versus random deployment, where n is the number of sensors needed [KUM 04]. Even if many optimizations can be done once the WSN is deployed to enhance its performance (e.g. sensor activity scheduling), we believe that the most important optimizations are those done at the design step to build the best topology possible that meets specific user requirements. This will be discussed in the next chapter.

3

Deterministic Deployment

Wireless sensor network (WSN) performance is directly related to the placement of the sensors within the region of interest. This chapter investigates the static WSN deterministic deployment, which aims at generating a network topology that satisfies user's requirements. It highlights the components involved and discusses the existing literature. Moreover, it analyzes the uncertainty-aware WSN deployment where sensors may not always provide reliable information and shows how the evidence theory could be exploited to design better deployment strategies. A comprehensive methodology for deterministic deployment of WSNs is presented and executed to deploy a simplified indoor surveillance WSN for motion detection.

3.1. Why deterministic deployment?

A worthy point here is that the WSNs deterministic deployment problem has different appellations in the literature, e.g. placement, layout, coverage, or positioning problems. The two major questions that need to be answered are: how many sensors should be deployed and where? In fact, the answers define several WSN properties such as coverage, connectivity and lifespan.

In deterministic deployment, the locations of sensors are precomputed prior to WSN start-up, which is usually pursued for indoor applications [AKY 02, YOU 08], when sensors are expensive, or when their operation is significantly affected by their position. In contrast to random deployment, deterministic deployment provides optimum network configuration, since positioning of sensors is determined beforehand to meet the design goals such as reducing cost and increasing coverage, connectivity, and lifetime. The most important metric considered in the literature is coverage. In fact, the research community has, in large part, investigated how to maximize the coverage rate while minimizing the usage of sensors (cost). The next section formalizes the deterministic deployment problem.

3.2. Formalization of the deterministic deployment problem

In the literature, it is generally assumed that events/targets appear at known locations referred to as target points defining the set $T \subseteq RoI$. T is determined according to user requirements. For instance, if full area coverage is required, target points can be selected uniformly and densely in the RoI. The deployment of sensors is constrained in the set $D \subseteq RoI$ consisting of deployment points in the RoI. These locations strongly depend on the specific application scenario [AMA 12]. Generally, they can be defined on the basis of a preliminary analysis or survey of the RoI, or the availability of a support or infrastructure (e.g. points where sensors can be easily hidden, points where they cannot be damaged, etc.). When sensors can be deployed anywhere in the RoI (i.e. $D = RoI$) and full area coverage is required (i.e. $T = RoI$), the RoI is discretized by a grid of points. Figure 3.1 illustrates a RoI discretized by a $m \times n$ grid wherein cells' centers are potential positions for sensors/targets.

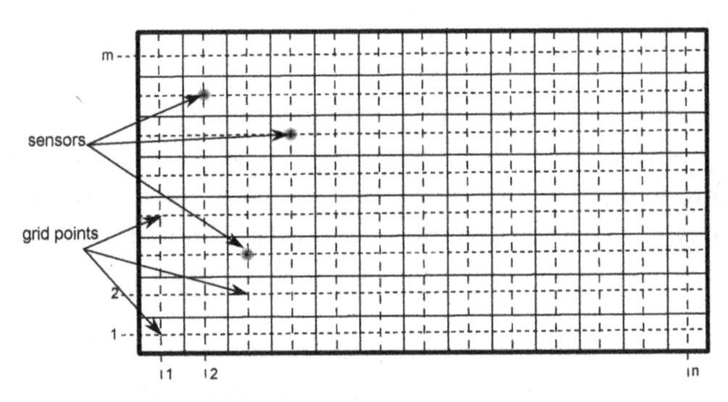

Figure 3.1. *Two-dimensional RoI model*

Regarding coverage requirements, it is considered that each target point $p \in T$ is associated with a required detection probability, denoted by R_p. The main objective of the WSNs deterministic deployment problem is to minimize the number of sensors while also satisfying the coverage requirements for all target points. Formally, the sensor placement problem can be formalized as follows:

$$\min \sum_{p \in D} x_p \qquad [3.1]$$

$$\text{s.t.} \quad \mathbb{P}_p \geq R_p, \ \forall p \in T \qquad [3.2]$$

$$x_p = 0 \ or \ 1, \ \forall p \in D \qquad [3.3]$$

Equation [3.1] defines the optimization problem where the objective is to minimize the deployment cost. The solution is constrained in equation [3.2], which requires that each target point is covered, i.e. $\forall p \in T$, the generated event detection probability at p (denoted by $\mathbb{P}_p$) must be equal to or greater than R_p. Equation [3.3] defines x_p as a zero-one variable. If $x_p = 1$, then a sensor will be deployed at the deployment point p. A point worth noting here is that the final value of $\mathbb{P}_p$ depends on the considered sensor coverage model. For instance, in the case of the evidence-based coverage model [SEN 12c], $\mathbb{P}_p = BetP(\{\theta_1\})_p$.

Other deployment-related issues could be included in the formulation above. For instance, to ensure network connectivity, the connectivity graph $G = (V, E)$ must be connected. V is the set of vertex formed by the deployed sensors, and E is the set of edges. Assuming a simple transmission disk model (see section 1.1.1.2), a unit disk graph can be used to describe the network. CC_G denotes the set of connected components of $G = (V, E)$. The cardinal of CC_G must be equal to 1 in order to guarantee the network connectivity. This problem can be formalized as follows:

$$\begin{aligned} &\min \sum_{p \in D} x_p \\ &\text{s.t.} \ \mathbb{P}_p \geq R_p, \ \forall p \in T \\ &\quad |CC_G| = 1 \\ &\quad x_p = 0 \ or \ 1, \ \forall p \in D \end{aligned}$$

The problem at hand is a binary integer programming problem which is NP-complete [PAP 81, KE 11]. To deal with such complexity, many heuristics have been proposed. Usually, the deterministic deployment of WSNs involves two components: (i) a *sensor coverage model* and (ii) a *placement algorithm*. By using a coverage model, a placement algorithm computes the minimum number of sensors along with their locations in order to achieve the desired design goals. The next section reviews the literature related to placement algorithms.

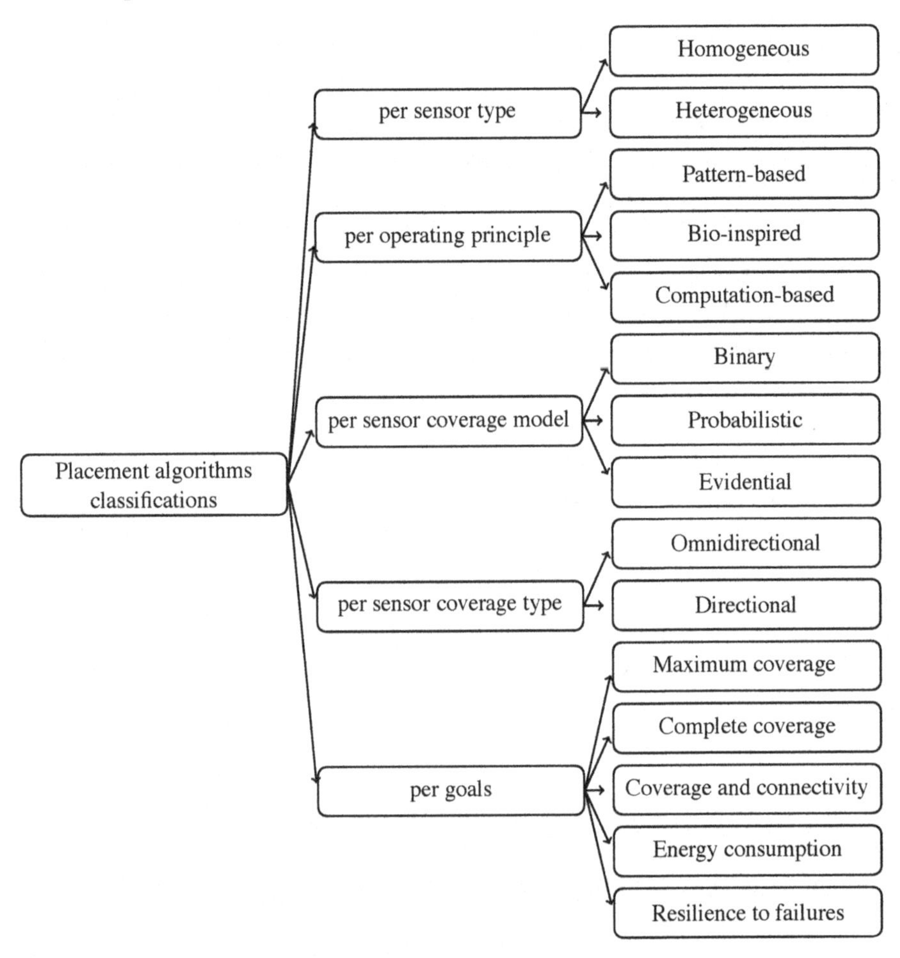

Figure 3.2. *High level taxonomy*

3.3. Placement algorithms

The analysis of WSN deployment approaches shows the existence of different classifications based on distinct goals. Figure 3.2 shows a high-level taxonomy for different possible classifications.

In this book, we classify the published strategies based on the used sensor coverage model. Thus, these strategies can be classified into three groups: (i) binary deployment strategies, (ii) probabilistic deployment strategies, and (iii) evidential deployment strategies. A taxonomy for deterministic deployment strategies is depicted in Figure 3.3.

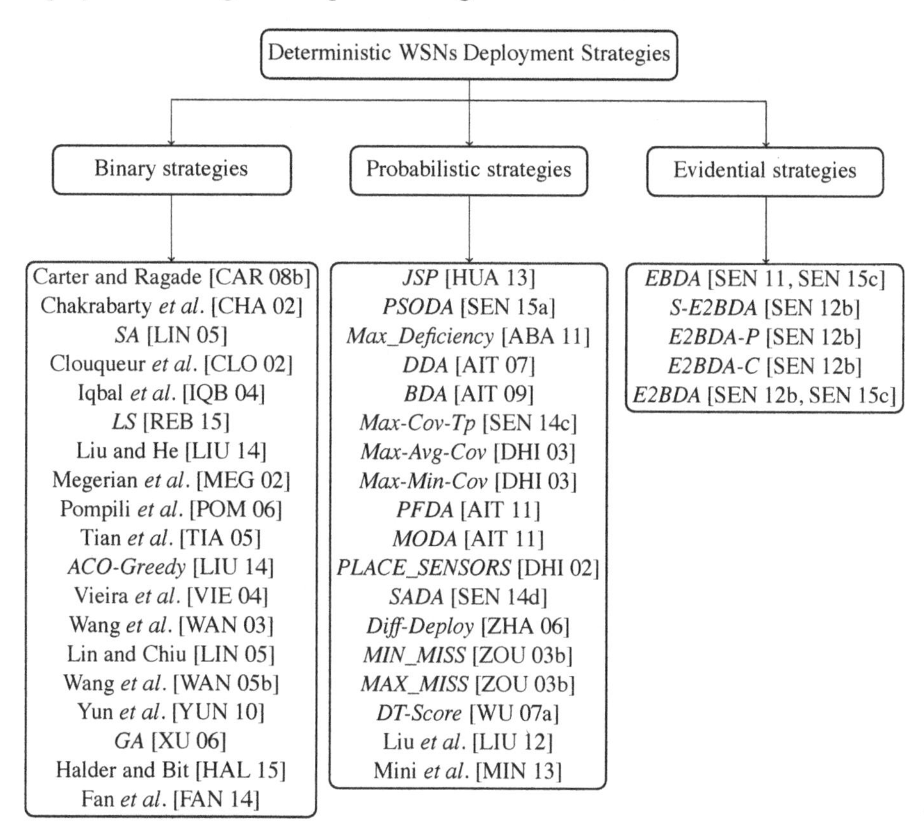

Figure 3.3. *Taxonomy for deterministic deployment strategies*

3.3.1. *Binary deployment strategies*

We talk about binary deployment strategies when the used sensor coverage model is the binary model (see section 1.1.1.1). When considering a binary coverage model, the sensor deployment problem can be formulated as the famous art gallery problem (AGP) addressed by the art gallery theorem

[ORO 87]. The AGP seeks to determine the minimum number of cameras that can be placed in a polygonal environment such that every point in the environment is monitored. Similarly, the deployment problem basically deals with placing a minimum number of sensors such that every point in the RoI is covered. The AGP has been solved optimally in two-dimensions and shown to be NP-hard in the 3-dimensional case [ORO 87]. Several variants of the AGP have been studied in the literature, including mobile guards, exterior visibility, and polygons with holes.

How to find the optimal placement pattern to fully cover a plane has been widely studied in the literature [ZHU 15]. In 1939, Kershner [KER 39] proved that the regular triangular lattice is the optimal placement pattern to blanket-cover an unbounded plane. In a regular deployment (also called pattern-based deployment), sensor-positions follow a regular topology, e.g. hexagonal, ring, star, etc. Many papers have studied regular deployment and compared it with constant diffusion (see section 2.2.2.1). In [MEG 02], the authors proved that the quality of monitoring in regular deterministic deployment is better than constant diffusion. A comparison was made with different regular topologies such as triangle, square, and hexagon. In [WAN 05b], the authors proposed to deploy the sensors according to a grid topology. In [POM 06], the authors investigated the problem of achieving maximal coverage with the lowest sensor count in the context of underwater WSNs, and proposed a triangular grid topology. In [VIE 04], an incremental deployment process was proposed. The main idea was to deploy a new sensor s at a point p, if no sensors are placed inside the sensing disk of s. According to [IQB 04], the proposed deployment process builds a hexagon-structured topology. The authors showed that the hexagon topology obtains better performances in terms of the minimum number of sensors required when compared with the triangular and square topologies. Authors in [CAR 08b] try to minimize costs while covering all target points. Yun *et al.* [YUN 10] proposed a set of deployment patterns to achieve full area coverage and k-connectivity ($k \leq 6$). Authors in [HAL 15] proposed a system based on Archimedes' spiral pattern that aims at extending the WSN lifetime. Zhu and Wang [ZHU 15] studied the optimal placement pattern in terms of the smallest sensor density. They prove that the regular triangular lattice is the optimal placement pattern among all the placement patterns consisting of regular polygons. In [FAN 14], the authors employed the regular hexagonal cell architecture to build a two-tier WSN consisting of regular sensors and relay nodes, while ensuring the minimum number of sensors under the constraints of coverage and connectivity. Authors in [AMM 14] discussed regular polyhedron patterns such as octahedron, cube, and tetrahedron, which are used in 3D WSNs.

In [CLO 02], the authors used a sequential deployment of sensors, i.e. a limited number of sensors is deployed in each step until the desired minimum exposure or probability of detection of a target is achieved. In most practical applications, however, we need to deploy the sensors in advance without any prior knowledge of the target. Therefore, sequential deployment is often infeasible. Moreover, sequential deployment may be undesirable when the number of sensors or the RoI is large. Thus, a single-step deployment scheme is more advantageous in such scenarios.

Sensor placement on two and three-dimensional grids was formulated as a combinatorial optimization problem and shown to be NP-complete [KE 11]. This approach has two main drawbacks. First, computational complexity makes the approach infeasible for large problem instances. Second, the grid coverage approach relies on "perfect" sensor detection, i.e. a sensor is expected to yield a binary yes/no detection outcome in every case. The grid-based WSNs deployment problem has been studied extensively and several approaches have been designed to solve it, such as integer linear programming [CHA 02], local search algorithm [REB 15], simulated annealing [LIN 05], genetic algorithm [XU 06, REB 15], and ant colony optimization [LIU 14].

When dealing with binary deployment strategies it should be noted that there is a relationship between connectivity and coverage. To the best of our knowledge, Wang *et al.* [WAN 03] and Zhang *et al.* [ZHA 05] are the earliest papers that independently proved the same conclusion: if a convex RoI is completely covered by a set of sensors, the communication graph consisting of these sensors is connected when $R_C \geq 2R_S$. In other words, under the condition that $R_C \geq 2R_S$, a WSN only needs to be configured to guarantee coverage in order to satisfy both coverage and connectivity. Hence, if the transmission range R_C of a sensor is much longer than its sensing range R_S then connectivity is not an issue. This result has been generalized for the case of k-coverage [TIA 05]. In short, k-coverage also implies k-connectivity when $R_C \geq 2R_S$.

3.3.2. *Probabilistic deployment strategies*

In probabilistic deployment strategies, the sensor detection is modeled probabilistically. In [DHI 02, DHI 03, ZOU 03b], authors considered a probabilistic coverage model. The RoI is assumed as a two-dimensional $n \times n$ grid of points. Dhillon *et al.* [DHI 02] formulated the sensor placement as an optimization problem. They proposed a greedy heuristic, called *PLACE_SENSORS*, to minimize the number of sensors and determine their locations to ensure full coverage of the RoI. *PLACE_SENSORS* complexity is equal to $O(n^4)$. The same authors proposed two placement heuristics in [DHI 03] also. In the first one, *Max-Avg-Cov*, the objective is to maximize the mean coverage of the grid points. In the second one, *Max-Min-Cov*, the goal is to maximize the coverage of the grid points that are the least effectively covered. The computational complexity of both algorithms is $O(n^4)$.

Dhillon *et al.* [DHI 02, DHI 03] defined four main variables D, M, M^* and M_{min}. $D = [P_{i/j}]$ is a sensor detection matrix for all pairs of grid points in the RoI. For an $n \times n$ RoI, we have a total of n^2 grid points, hence the matrix D consists of n^2 rows and n^2 columns, and a total of n^4 elements. $M = m_{ij} = 1 - P_{i/j}$ is a miss probability matrix. The vector $M^* = (M_1, M_2, ..., M_N)$ is the set of miss probabilities for the $N = n^2$ grid points in the RoI. At the start of the placement algorithm, no sensor is deployed so M^* is initialized to the all-1 vector, i.e. $M^* = (1, 1, ..., 1)$. Each sensor placed in the RoI decreases one or more entries in this vector. M_{min} is the maximum value of the miss probability that is permitted for any grid point. In *Max-Avg-Cov* a sensor is deployed at a grid point i where $\sum_j M_{ij}$ is minimal. However, in *Max-Min-Cov* a sensor is deployed at the grid point of minimum coverage in the RoI. The placement algorithms are stopped when the generated detection probabilities are greater than the requested detection probabilities, or the number of sensors deployed exceeds the allocated budget for sensors. The pseudo code of *Max-Avg-Cov* (resp. *Max-Min-Cov*) is outlined in algorithm 3.1 (resp. algorithm 3.2).

The computational complexity for both *Max-Avg-Cov* and *Max-Min-Cov* is $O(n^4)$. In addition, both algorithms create a matrix that contains n^4 elements, so the memory complexity for both *Max-Avg-Cov* and *Max-Min-Cov* is $O(n^4)$. A similar method, known as the jigsaw-based sensor placement (JSP) algorithm, was proposed in [HUA 13]. JSP starts by placing the sensors at the periphery of the RoI. After that, sensors are deployed in the RoI in an iterative manner while avoid dividing the uncovered area into isolated regions. The computational complexity of JSP is $O(n^2)$. In [LIU 12], the authors present an immune-swarm intelligence (ISI)-based sensor placement algorithm that combines the particle swarm optimization technique (PSO) and artificial immune systems.

Algorithm 3.1. Max-Avg-Cov (M, M^*, M_{min})

1: $num\_sensors = 1$;
2: **repeat**
3: **for** $i = 1$ to N **do**
4: $\sum_i = m_{i1} + m_{i2} + ... + m_{iN}$;
5: **end for**
6: Place sensor on grid point k such that $\sum_k$ is minimum;
7: **for** $i = 1$ to N **do**
8: $M_i = M_i m_{ki}$;
9: **end for**
10: Delete k^{th} row and column from the M matrix;
11: $num\_sensors = num\_sensors + 1$;
12: **until** $(M_i < M_{min}$ for all $i, 1 \leq i \leq N) \vee (num\_sensors > limit)$

Algorithm 3.2. Max-Min-Cov (M, M^*, M_{min})

1: Place first sensor randomly;
2: $num\_sensors = 1$;
3: **repeat**
4: **for** $i = 1$ to N **do**
5: $M_i = M_i m_{ki}$;
6: **end for**
7: Place sensor at grid point k such that M_k is max;
8: Delete k^{th} row and column from the M matrix;
9: $num\_sensors = num\_sensors + 1$;
10: **until** $(M_i < M_{min}$ for all $i, 1 \leq i \leq N) \vee (num\_sensors > limit)$

Wu *et al.* [WU 07a] looked at the deployment problem from a geometric angle. They proposed *DT-Score*, a two-step placement algorithm. In the first step, *DT-Score* attempted to heal the coverage holes close to the boundary of the RoI and obstacles by placing sensors near them. The second step exploits the Delaunay triangulation to determine the sensors' locations. The computational complexity of *DT-Score* is $O(n^2 \log n)$.

Zou and Chakrabarty [ZOU 03b] examined the uncertainty related to the precomputed sensor positions. The authors proposed to model sensor positions as a random variable with a Gaussian probability distribution, and devise two placement heuristics: *MIN_MISS* and *MAX_MISS*. Both heuristics are iterative, where in each iteration, only one sensor is deployed. In *MIN_MISS*, a sensor is placed at the grid point with the lowest over miss probability, whereas, in *MAX_MISS*, a sensor is placed at the grid point that maximizes the over-miss probability. *MIN_MISS* and *MAX_MISS* computational complexity is equal to $O(n^6)$.

The aforementioned deployment strategies addressed the uniform deployment problem in which the monitored event has the same importance at every point of the RoI. This means that the requested detection probabilities for the entire RoI are identical. In [ZHA 06, AIT 07, ABA 09b, AIT 09, AIT 11, SEN 14c, SEN 14d, SEN 15a], the authors explicitly addressed the non-uniform deployment problem, in which the required detection probabilities at different positions are not necessarily identical. Aitsaadi *et al.* [AIT 07] exploited mesh representation techniques used in image processing, and devised a placement algorithm called *DDA*. This latter suffers from a high computational complexity $O(n^6)$. The same authors presented a placement algorithm, *BDA* [AIT 09], based on the Tabu search metaheuristic. *BDA* has a quadratic complexity, but, in spite of the number of iterations, does not ensure full area coverage. In [SEN 14c], Senouci *et al.* devised a greedy heuristic called Max-Cov-Tp. The main advantage of the *Max-Cov-Tp* algorithm consists of avoiding the maximization of the detection probability of target points that have been already covered which reduces the deployment cost. The computational complexity of *Max-Cov-Tp* is $O(n^6)$. The same authors proposed in [SEN 14d] a polynomial-time deployment algorithm called simulated annealing-based sensor deployment algorithm (*SADA*). The obtained results show that *SADA* algorithm provides better results than other heuristics such as *MIN_MISS* and *BDA*.

Authors in [MIN 13] assumed a limited number of sensors and employed an artificial bee colony (ABC) algorithm to compute the optimal deployment locations of sensors that satisfies the different targets coverage requirements. In [SEN 15a], Senouci *et al.* introduced *PSODA*, an improved binary PSO-based polynomial-time sensor placement algorithm. The obtained results showed that *PSODA* outperforms the state-of-the-art approaches, especially in the case of preferential coverage.

Authors in [ZHA 06] formalized the deployment problem as a linear shift-invariant (LSI) system, and proposed a heuristic called *Diff-Deploy*. This latter exhibited a high computational complexity equal to $O(\frac{4}{3}n^6)$. Ababnah *et al.* [ABA 09b, ABA 11] formalized the deployment problem as a linear quadratic regulator while considering a fixed number of available sensors. As the optimal control-based solution suffered from a very high complexity, Ababnah *et al.* [ABA 09b, ABA 11] proposed an approximation heuristic called *Max_Deficiency* whose complexity is $O(n^6)$.

Table 3.1 provides a comparative summary of the various probabilistic deployment approaches discussed above.

3.3.3. *Evidential deployment strategies*

When considering an evidential coverage model (see section 1.1.1.1), the WSN deployment problem has been formalized as a binary nonlinear and non-convex optimization problem [SEN 11, SEN 12b].

Considering only uniform coverage and cost objectives, Senouci *et al.* [SEN 11] presented *EBDA*, an evidence-based deployment algorithm that considers a grid-like deployment. EBDA is an incremental deployment algorithm that determines the locations of the sensors one at a time. In each step, it finds all possible locations that are available on the grid for a sensor, and calculates the collective coverage (denoted by *CC*) associated to this sensor in addition to those already deployed. A sensor is placed at the position that allows the maximum number of detection decisions. When the best location is found for the current sensor, the *CC* is updated to include the newly deployed sensor. As the objective is to ensure that every target point is covered, this intended outcome is the first termination criterion. The number of available sensors (NAS) is defined as a second termination criterion. The pseudo-code of EBDA is shown in Algorithm 3.3.

Strategy	*Computational complexity*	*Memory complexity*	*Primary objective*	*Secondary objective*	*Constraint*
PLACE_SENSORS [DHI 02]	High	High	Max. the average coverage	Min. the number of sensors	-
Max-Avg-Cov [DHI 03]	High	High	Max. the average coverage	Min. the number of sensors	-
Max-Min-Cov [DHI 03]	High	High	Min. miss detection proba	Min. the number of sensors	-
JSP [HUA 13]	Medium	Medium	Eliminates isolated regions	Min. the number of sensors	-
Mini *et al.* [MIN 13]	Medium	Medium	Satisfies uniform coverage	Min. the number of sensors	-
DT-Score [WU 07a]	Medium	Medium	Eliminates coverage holes	Min. the number of sensors	-
MIN_MISS [ZOU 03b]	Very high	Medium	Min. the over miss proba	Min. the number of sensors	Locations uncertainty
PFDA [AIT 11]	Medium	Medium	Satisfies preferential coverage	Min. the number of sensors	Network connectivity
MODA [AIT 11]	High	Medium	Satisfies preferential coverage	Min. the number of sensors	Connectivity and lifetime
MAX_MISS [ZOU 03b]	Very high	Medium	Max. the over miss proba	Min. the number of sensors	Locations uncertainty
DDA [AIT 07]	Very high	Medium	Satisfies preferential coverage	Min. the number of sensors	-
Liu *et al.* [LIU 12]	Medium	Medium	Satisfies Q-coverage	-	Fixed number of sensors
Max-Cov-Tp [SEN 14c]	Very high	Medium	Max. covered points	Min. the number of sensors	-
BDA [AIT 09]	High	Medium	Max. preferential coverage	Min. the number of sensors	-
Diff-Deploy [ZHA 06]	Very high	High	Satisfies preferential coverage	Min. the number of sensors	-
SADA [SEN 14d]	High	Medium	Satisfies preferential coverage	Min. the number of sensors	-
Max_Deficiency [ABA 11]	Very high	Medium	Satisfies preferential coverage	-	Fixed number of sensors
PSODA [SEN 15a]	Medium	Medium	Satisfies preferential coverage	Min. the number of sensors	-

Table 3.1. *A comparison between the various placement algorithms*

Algorithm 3.3. EBDA

Require: $RoI, S\ (set\ of\ sensors), Th, NAS$

Ensure: PR ($PlacementResult$), U

1: Set $CC = [0]$, $PR = [0]$, $U = \{\emptyset\}$
2: **for** (every target point $p \in T$) **do**
3: Compute SC_p *//coverage generated by a sensor located at point p*
4: **end for**
5: **repeat**
6: **for** (every deployment point $p \in D$) **do**
7: Evaluate CC
8: Place sensor $s \in S$ on deployment point p such that CC is maximum
9: Set $S = S - \{s\}$, $U = U \cup \{s\}$
10: Update PR
11: **end for**
12: **until** (RoI is covered using PR) **or** ($|U| > NAS$)

The computational complexity of EBDA is equal to $O(n.m)$ where $n \times m$ is the size of the RoI. The convergence and the effectiveness of EBDA are reported in [SEN 11, SEN 15c].

In [SEN 12b], Senouci *et al.* considered two additional constraints, namely preferential coverage and network connectivity. They characterized the structure of an optimal solution that can be used to construct an optimal solution to the problem from optimal solutions to subproblems. However for this deployment problem, the total number of distinct subproblems is an exponential in the input size. This will lead us to an exponential-time algorithm. As an alternative, they proposed a polynomial-time algorithm efficient evidence-based sensor deployment algorithm (*E2BDA)* to construct a suboptimal solution.

Suppose that a placement strategy of n sensors $s_1, ..., s_n$ is D_n. A target point p is covered if $\mathbb{P}_p \geq R_p$. Let us denote $score_{D_n}$ the number of coverage decisions reached by a fusion center using evidence from the set of deployed sensors $s_1, ..., s_n$ in the RoI according to the placement strategy D_n. The deployment of an additional sensor s_{n+1}, generated a set of possible placement strategies D^q_{n+1}, where q corresponded to the location of the newly deployed sensor s_{n+1}. The obtained $score_{D^q_{n+1}}$ can be regrouped in a matrix (denote by PS).

Initially, the placement of one sensor s_1 in the RoI is considered. For all possible locations on the grid, the algorithm computes the associated PS matrices and saves the k-best placements of s_1 denoted by BP. k is considered as an input parameter, which can take any positive value. After that, in each step i, the algorithm finds all possible locations that are available on the grid for s_i, and calculates the matrices PS associated to s_i in addition to the sensors already deployed $s_1, ..., s_{i-1}$. BP are updated to save the k-best placements found for $s_1, ..., s_i$. In the last step, the algorithm selects the best placement over the k-best placements BP found for $s_1, ..., s_n$, where $n \leq NAS$. The E2BDA pseudo-code is illustrated in algorithm 3.4.

It is important to point out that the vacuous belief function is a neutral element with respect to the combination operation, E2BDA does not check all grid points to compute the achieved coverage but only a limited zone using a sliding window-based technique. If NAS is large, the computational complexity of E2BDA is equal to $O(kw^2mn)$, where k is the number of the best placements considered at each iteration, and w is the window size which is equal to the sensor's sensing rang R_s. The product kw^2 is equal to a constant C, hence the computational complexity of E2BDA is equal to $O(Cmn)$. E2BDA has been compared to the other deployment strategies such as: Max-Avg-Cov [DHI 03], Max-Min-Cov [DHI 03], MIN_MISS [ZOU 03b], and EBDA [SEN 11]. Experimental results reported in [SEN 12b, SEN 15c] show that E2BDA outperforms these deployment strategies. In the next section, we will discuss how such deterministic deployment strategies could be used to deploy a real-world WSN.

3.4. A simple methodology for deterministic deployment of WSNs

The main idea of deterministic deployment approaches is the determination of the positioning of sensors beforehand to meet the design goals. In other words, we try to estimate, by simulations, the WSN performances before its deployment and we build the best possible topology that meets specific user requirements. The real-world WSN will be deployed as suggested by the obtained topology. The fundamental problem here is to ensure that the deployed WSN meets the design goals.

Algorithm 3.4. E2BDA

Require: RoI, S, Th, k, NAS, R_c
Ensure: $optimalPlacement, deployedSensors$
{Initialization}
1: $deployedSensors = 0$
2: Set $BP = [0]$
3: **for** (every target point $p \in RoI$ such as $Th_p > 0$) **do**
4: Compute SC_p
5: **end for**
{Sensors placement}
6: **repeat**
7: Using BP and R_c compute $U = \{p_{ac}, p_{ac} \in RoI$ and $Th_{p_{ac}} > 0$ and p_{ac} *available* and $p_{ac} \in$ *the connectivity area of BP* $\}$
8: **for** (every grid point $p_{ac} \in U$) **do**
9: Compute the associated PS matrices using BP, SC_{p_a} and Th
10: **end for**
11: Update BP to save the k-best placements
12: $deployedSensors = deployedSensors + 1$
13: $optimalPlacement$ = the best placement in BP
14: **until** (RoI is covered using $optimalPlacement$) **or** ($deployedSensors > NAS$)

Not much work has been done to devise a comprehensive methodology for deterministic deployment of wireless sensor networks. For instance, it is still not very clear how to deploy a WSN that achieves the design goals. Furthermore, most of the current work provides placement algorithms with no experimental validation. To address this issue, Senouci *et al.* [SEN 14b] devised a simple methodology to guarantee that the deployed WSN meets the design goals. The proposed methodology can also be exploited to further analyze, compare, and validate deterministic WSNs deployment approaches.

When using simulations to estimate the WSN performances, it is worth noting that the quality of this estimation is determines by far the real quality of service of the deployed WSN, since the choice of the best topology is guided by this estimation. In addition, the quality of this estimation depends mainly on the used abstract models. In fact, in the estimation process, the key components are the abstract models. For instance, if we are dealing with the issues of sensing coverage and network connectivity, then the abstract models will include a sensor coverage model and a sensor communication model.

The discussion above suggests that to ensure that the deployed WSN meets the design goals, we need to guarantee a good estimation of the WSN performances initially. This can be achieved by building the abstract models carefully. Once models are built, the methodology uses scenario driven simulations to rank deployment alternatives based on the deployment goals. Finally, real measurements are carried out to verify that the WSN meets the design goals. Figure 3.4 shows an overview of this methodology.

The proposed methodology [SEN 14b] consists of the following steps:

– *Step 1:* Build good abstract models;

- build a good abstract sensor coverage model by estimating the real sensing performances of the considered sensors,

- build a good abstract sensor connectivity model by examining the basic characteristics of the communication channel,

- if other issues are considered such as the network lifetime, then build the corresponding abstract model;

– *Step 2:* Using the abstract models built in Step 1, a network model and a placement algorithm, build, through simulations, the best possible topology that meets specific user requirements. Likewise, rank deployment alternatives if required;

– *Step 3:* Deploy the real-world WSN as suggested by the topology obtained in Step 2;

– *Step 4:* Measure the real performance of the WSN deployed in Step 3, if it doesn't meet the design goals goto Step 1.

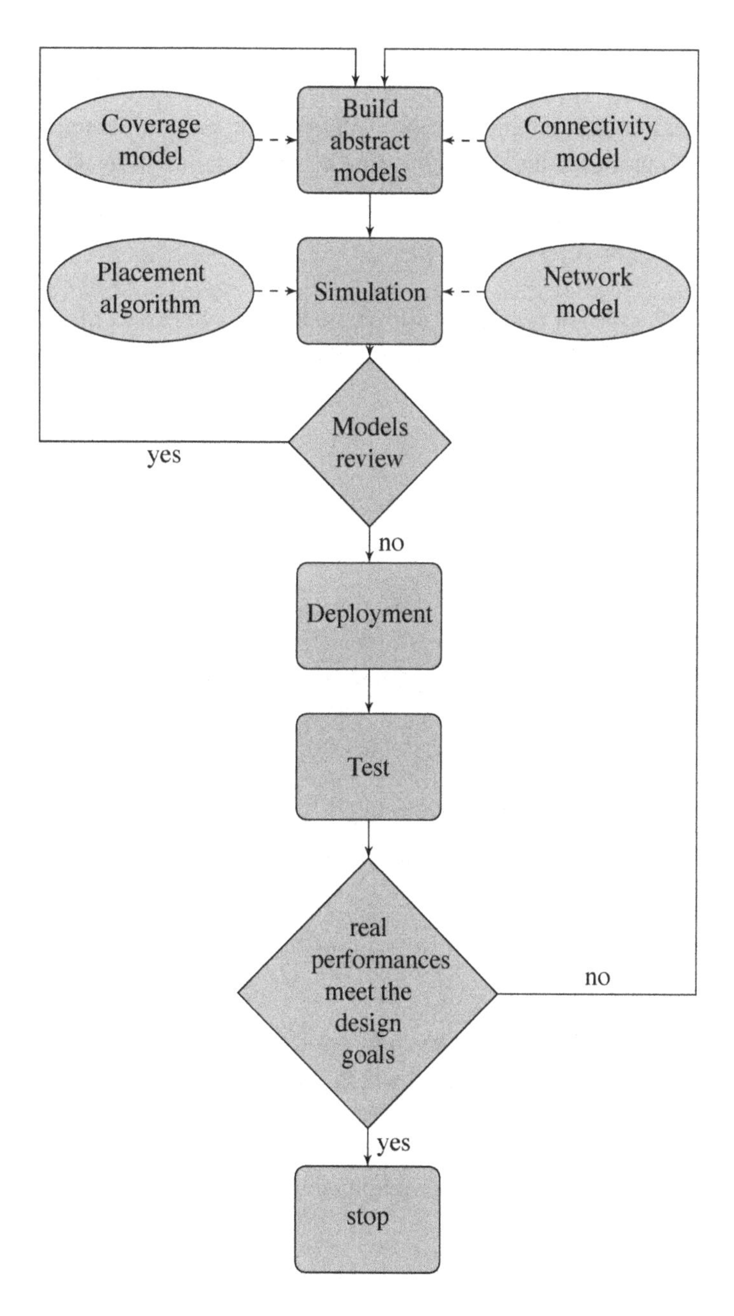

Figure 3.4. *Methodology overview*

3.5. Application: deploying a surveillance WSN

The proposed deployment methodology was executed to deploy a simplified indoor surveillance WSN, *ArduiNet*, for motion detection. In this section, we summarize the deployment process and the main results.

3.5.1. *Hardware and software*

Figure 3.5 shows a typical ArduiNet node. It's composed of an Arduino UNO board, a PIR Phidgets 1111_0 sensor, and IEEE 802.15.4 radio (XBee module).

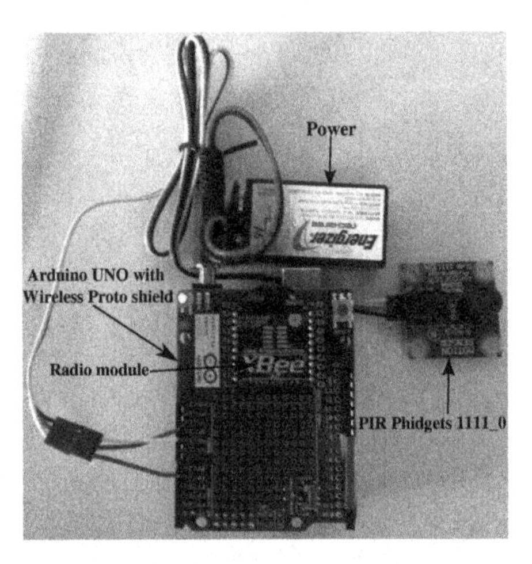

Figure 3.5. *Typical ArduiNet node. For a color version of the figure, see www.iste.co.uk/senouci/wireless.zip*

The PIR sensors detect changes in infrared radiation that occur when there is movement by a person (or object) which is different in temperature from the surroundings. It should be noted that the PIR Phidgets sensor has an analog output. Regarding the software, the Arduino IDE [ARD 15] is used to create a sketch (a computer program) that we upload to the Arduino board.

3.5.2. *System architecture*

ArduiNet was deployed in a testbed room as shown in Figure 3.6. A 2.4 m × 4 m RoI is discretized by a two-dimensional grid of points. The granularity of the grid, i.e. the distance between consecutive grid points, is 0.4 m. Each cell's center is a possible location for a sensor as illustrated in Figure 3.6.

Figure 3.6. *System architecture. For a color version of the figure, see www.iste.co.uk/senouci/wireless.zip*

3.5.3. *Experiments and results*

In this section, we report the obtained results when comparing the probabilistic-based approach and the evidence-based approach. First, we discuss how to build belief functions from raw sensory data of the PIR Phidgets 1111_0 sensor. Second, we examine the basic characteristics of the IEEE 802.15.4 communication channel. Third, we analyze the performance of ArduiNet as a surveillance system while using a probabilistic-based approach and the evidence-based approach.

3.5.3.1. *Coverage model analysis*

Figure 3.7 gives an example of a set of mass functions associated to the PIR Phidgets sensor in an indoor environment. When there is no movement, this sensor provides a value around 500 (value gauged by the manufacturer). Any other measure is equivalent to its symmetrical around 500. Numbers outside the 400–600 range denote the detection of a moving object with a high certainty. A projection on the set of mass functions is done in order to obtain the corresponding mass function each time raw data is received from that sensor. For instance, if the Phidgets sensor returns a value of 100, then the resulting mass function would have two focal elements: $m(\{\theta_1\}) = 0.95$ and $m(\Theta) = 0.05$.

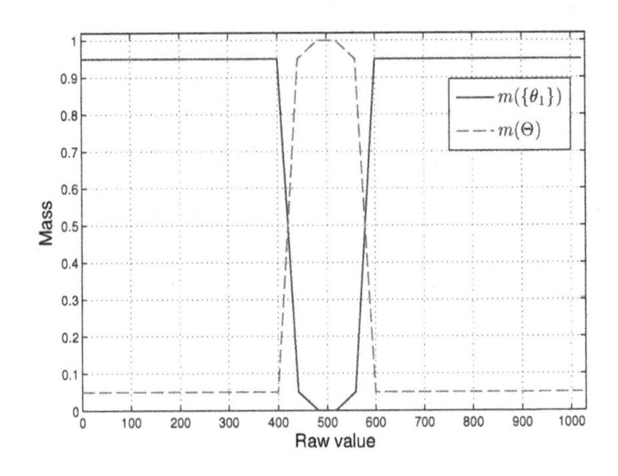

Figure 3.7. *Example of a set of mass functions associated to the PIR Phidgets 1111_0 sensor. For a color version of the figure, see www.iste.co.uk/senouci/wireless.zip*

3.5.3.2. *Connectivity model analysis*

In this section, we evaluate through measurements, using the XBee module (IEEE 802.15.4 radio), the Packet Reception Ratio (PRR) in the testbed environment. This study will allow us to correctly calibrate the connectivity model which is an input to the deployment framework.

Since the distance and the direction are fundamental factors that affect the link quality, in this experiment, we place the transmitter at different distances and directions from the receiver, and compute the PRR. We set the packet size to 11 bytes. The obtained results are reported on Figure 3.8. Note that the reported PRR is the average over different PRR samples.

Figure 3.8 clearly shows three distinct reception regions: connected, transitional, and disconnected. This confirms what has been revealed by many recent studies such as [PET 06]. The connected region is characterized by high reception rates, which means that the links have good connectivity. Since the connected region is much longer than our testbed room, connectivity is not an issue. In other words, our WSN testbed only needs to be configured to guarantee coverage in order to satisfy both coverage and connectivity.

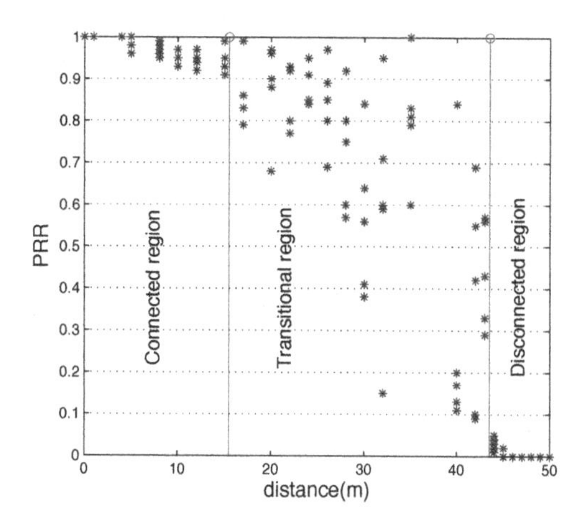

Figure 3.8. *Regions of connectivity. For a color version of the figure, see www.iste.co.uk/senouci/wireless.zip*

3.5.3.3. *Evidential deployment vs. probabilistic deployment*

We are interested in the ability of the ArduiNet system to detect walking persons, as demanded by the application. As we focus on quantifying the benefit of the evidence-based deployment approach rather than the efficiency of the placement algorithm, we have compared EBDA to MIN_MISS. The choice of these two algorithms is based on the following reasons: (i) both algorithms are iterative, where in each iteration, only one sensor is deployed, (ii) EBDA deploys a new sensor at a deployment point p that maximizes the number of detection decisions, while MIN_MISS deploys a new sensor at a deployment point p that minimizes the over miss probability value, (iii) EBDA exploits the evidence-based coverage model, while MIN_MISS is based on the probabilistic coverage model. This choice allows us to quantify the real benefit of the evidence-based approach while minimizing the differences due to the efficiency of the placement algorithms. In individual tests, we deploy the ArduiNet system as suggested by the output of the considered placement algorithms (EBDA and MIN_MISS), we let a single person walk by at different paths in the test bed room, and we measure the detection rate. The experiments were repeated more than 1,000 times. The obtained results are summarized in Table 3.2.

	Requested detection rate %	*Achieved detection rate %*	*Deployment cost*
EBDA	90	91	6
	95	97	8
MIN_MISS	90	94	8
	95	98	11

Table 3.2. *Experimental results*

We see clearly from Table 3.2 that fewer sensors are required when EBDA is used. EBDA consistently outperforms MIN_MISS in all cases. In fact, while both strategies guarantee the minimum required detection probability, the evidence-based approach saves almost 25% of the deployment cost over the probabilistic one. As mentioned in [SEN 15c], the rate of reduction in the deployment cost achieved by the evidence-based approach is related among others to the size of the RoI and the minimum required detection probability. For instance, in a 2.4 m × 4 m RoI, the evidence-based approach saves 25% (resp. 27%) of the deployment cost over the probabilistic one for a 0.9 (resp. 0.95) minimum required detection probability. Also, the larger the RoI, the greater the performance gain. These experimental results confirm that evidence combination improves sensing coverage significantly by exploiting the collaboration among sensors.

3.6. Practical issues that need further research

Though much work has been done in the field of deterministic deployment of WSNs, there is still a vast scope for future research in this area. Handling sensor reliability and environmental factors while considering the application needs is a major design challenge that needs to be addressed. Furthermore, most of the present works make unrealistic assumptions regarding the modeling of the environment and/or the sensors. Developing deterministic deployment approaches that address practical constraints is a major open problem in this field. The remainder of this section highlights the major challenges and open problems for future research work.

3.6.1. *Handling perturbations*

All real-life large-scale deployments strategies result in some randomness [BAL 09]. Some prime sources are sensors' reliability, placement errors, and environmental factors, either at the time of deployment or afterward.

By design, sensors could be vulnerable to misreading or malfunctioning due to their quality [ELO 04]. This intrinsic property must be regarded as a normal property of the sensor and should be taken into account prior to the WSN start-up. One possible approach was described by Senouci *et al.* [SEN 12d]. The authors consider sensors as only partially reliable. To anticipate this behavior (i.e. sensors unreliability), Senouci *et al.* propose to use the *a priori* knowledge on the reliability of a sensor in order to give less weight to the information provided by that sensor at the pre-deployment stage, i.e. to discount the sensor's beliefs. Zou and Chakrabarty [ZOU 03b] examined placement errors and proposed to model sensor positions as a random variable with a Gaussian probability distribution. Balister and Kumar [BAL 09] considered the effects of placement errors and random failures on the density needed to achieve full coverage while considering both random and deterministic deployment.

Although sensors may be of good quality and provide accurate readings, external factors can greatly influence their correct functioning [ELO 04]. In this case, sensors are vulnerable to misreading or malfunctioning due to their locations in the RoI. Senouci *et al.* [SEN 12d] called such locations "challenging locations". So far, few researchers have tried to deal with this issue in the WSN deployment process. In [FEI 09], the authors try to handle environmental factors such as obstacles and weather conditions at the post-deployment stage. They assume that sensors know the position of gathered event, by deploying samples in the network. After gathering samples, the sensing range of sensors are determined by a revised geometry structure (α-shape). Different from this work, Senouci *et al.* [SEN 12d] consider that a fully reliable sensor is considered as only partially reliable if it is deployed in a challenging location. They proposed to discount the confidence we had toward the sensors deployed in challenging locations, at the pre-deployment stage, using discounting factors associated with deployment points.

3.6.2. *Deterministic deployment of three-dimensional WSNs*

Most deterministic schemes only consider two-dimensional environments and ignore the effects of elevation and obstacles. However, with the emergence of three-dimensional (3D) applications, such as smart homes and smart cities, tackling the deterministic deployment issues in 3D has become a necessity. In this context, Topcuoglu *et al.* [TOP 11] have proposed a new formulation for the WSN deployment in a synthetically generated 3D environment. Although the proposed approach considers a realistic spatial model of the environment, it assumes a binary sensor coverage model. In [TEM 14], the authors use wavelet transformation to determine the coverage cavities, and propose a deployment algorithm that is based on the cat swarm optimization (CSO). Akbarzadeh *et al.* [AKB 13] proposed the integration of terrain information (elevation maps) with a probabilistic sensor coverage model. Future research in this area should make realistic assumptions regarding the modeling of the environment and the sensors (coverage, connectivity, etc.).

3.6.3. *Refocusing on application demands*

Usually, deployment of WSNs is application-specific. This may require the expertise of non-computer science-related fields. To explain the problem resulting from such cross-discipline applications, let consider the example of structural health monitoring. On one hand, existing civil engineering approaches do not seriously consider WSN constraints, such as communication load, network connectivity, and fault tolerance. On the other hand, the computer science community generally makes assumptions due to the lack of knowledge of civil engineering, which results in deployment approaches that do not satisfy civil engineering requirements. Consequently, both efforts are less useful from applications perspective. Recent efforts [LI 10, BHU 12, DU 15] have clearly shown that for the deployment of WSNs, cross-disciplined knowledge is possible and valuable. Thus, future work should study WSN deployment problem not only from a computer science efficiency perspective, but should also focus on application-specific requirements by emphasizing the collaboration between scientists from different disciplines.

3.7. Conclusion

This chapter focused on deterministic deployment strategies for static WSNs. The discussed approaches have been mainly concerned with determining the minimum number of sensors and their locations in order to satisfy detection requirements. The detection rule was a simple scheme in which a target or an event was detected if at least one sensor reports a detection decision. However, such a detection rule does not consider false alarm requirements. The expense of the latter should be considered especially in surveillance applications. In fact, most of all detection reports require some sort of physical response, so it can be very costly to respond to numerous false alarms over time. A viable approach to meet false alarm requirements is data fusion. In the next chapter, we would focus on deterministic deployment approaches that employ detection rules based on data fusion methods.

4

Fusion-based Deterministic Deployment

This chapter investigates the fusion-based deterministic deployment that is usually employed in the deployment of WSNs for critical applications that impose stringent requirements such as a high detection rate coupled with a low false alarm rate. The aim of fusion-based deterministic deployment is to produce a hierarchical network topology that satisfies the user's requirements. This chapter formalizes different variants of the fusion-based deterministic deployment problem, discusses existing sensor placement algorithms, shows how the evidence theory could be exploited to design better fusion-based deployment strategies, and formulates practical issues that need further research. As an example, this chapter reports the results obtained when deploying a real-world simplified fusion-based indoor surveillance WSN.

4.1. Why fusion-based deterministic deployment?

In mission-critical WSN surveillance applications, a high detection rate coupled with a low false alarm rate is essential. In other words, mission-critical WSNs surveillance applications require a good quality of coverage, not just a good coverage. In the previous chapter, discussed deployment approaches consider the coverage rate or detection rate as the main metric. Thus, they focus on the coverage rate and completely ignore the quality of coverage. Hence, generated topologies do not guarantee the false alarm rate. In contrast to the deployment approaches treated in the previous chapter, fusion-based deterministic deployment approaches consider both false alarm and detection requirements.

Critical applications, such as military target detection, impose stringent requirements for event detection accuracy. To detect a target, the sensors have to make local observations based on their surrounding environment and collaborate to produce a global decision that reflects the status of the region of interest (RoI). It is well-known that collaboration among sensors improves the sensing quality by jointly considering noisy measurements of multiple unreliable sensors. For example, He *et al.* [HE 06] show that the false alarm rate of a real-world Mica2 WSN can be reduced from 60% (when sensors decide independently) to near zero by adopting a fusion scheme.

4.2. Fusion-based detection coverage models

In contrast to the aforementioned coverage models (see section 1.1.1.1), fusion-based detection coverage models consider effective collaboration among sensors. This section reviews exiting fusion-based detection coverage models, which can be classified into two groups: (i) fusion-based probabilistic models, and (ii) fusion-based evidential models.

4.2.1. *Fusion-based probabilistic models*

An important application of WSNs is to detect events that could occur at some location. In the context of detection applications, the sensing quality of a sensor can be represented by its detection probability. The detection probability of a space point by a single sensor is related, among other factors, to the distance between them. However, the detection probability of a space point relative to a set of sensors is no longer computed as the summation of the detection probability of the point relative to each individual sensor (otherwise, it might be greater than one). Instead, a value fusion or a decision fusion scheme can be used to derive the detection probability.

In the value fusion scheme, the nodes exchange their raw energy measurements, whereas, in the decision fusion scheme, the nodes exchange local detection, decisions based on their energy measurement. In decision fusion, each sensor makes a local decision based on its measurements and sends it to the fusion center, which makes a system decision according to the local decisions. The optimal decision fusion rule has been obtained in [CHA 86]. In value fusion, each sensor sends its measurements to the fusion center, which makes the detection decision based on the received measurements. Value and decision fusion approaches were compared in [CLO 03, CLO 04]. In [CLO 03], it was found that the former performs better in terms of detection probability for low noise levels. However for noisy energy measurements, decision fusion proves more robust. Clouqueur *et al.* [CLO 04] show that decision fusion becomes superior to value fusion as the ratio of faulty sensors to fault free sensors increases. Furthermore, the communication cost is lower in decision fusion than in value fusion.

Two metrics are generally considered to characterize the performance of a detection system, namely: detection rate and false alarm rate. This latter (denoted by P_f) is the probability of making a positive detection decision when no target is present. Detection rate (denoted by P_d) is the probability that a target is correctly detected. Based on different event scenarios and detection techniques, many fusion-based detection coverage models have been proposed in the literature [WAN 07b, YAN 09, TAN 11, AMA 12]. Amaldi *et al.* [AMA 12] proposed a simple value fusion scheme. The authors consider the mobile target detection scenario, where sensing quality is based on the concept of path exposure. Yang and Qiao [YAN 09] studied the effects of sensor collaboration on barrier coverage. In [TAN 11], authors describe a value fusion scheme as follows: suppose sensor s is d_s meters from the target/event, the signal energy it measures is given by $U_s = W(d_s) + N_s{}^2$, where $W(d_s)$ denotes the signal energy measured by the sensor s, and N_s the noise strength. N_s follows the zero-mean normal distribution with a variance of σ^2, i.e. $N_s \sim \mathcal{N}(0, \sigma^2)$. Sensors send their energy measurements to the fusion center which, in turn, compares the average of all measurements against a threshold η to make a decision regarding the presence of the target/event. The threshold η is referred to as the detection threshold. In practice, the parameters of target and noise models are often estimated using a training dataset before deployment. As per the aforementioned value fusion scheme [TAN 11], the detection rate is given by:

$$P_d = \mathbb{P}\left(\frac{1}{n}\sum_{i=1}^{n}(W(d_i) + N_i^2) > \eta\right) = 1 - \mathbb{P}\left(\sum_{i=1}^{n}\left(\frac{N_i}{\sigma}\right)^2 \leq \frac{n\eta - \sum_{i=1}^{n} W(d_i)}{\sigma^2}\right)$$

As $N_s/\sigma \sim \mathcal{N}(0, 1)$, $\sum_{i=1}^{n}(N_i/\sigma)^2$ follows the Chi-square distribution with n degrees of freedom whose Cumulative Distribution Function (CDF) is denoted as $\chi_n(.)$. Hence, P_d can be calculated by:

$$P_d = 1 - \chi_n\left(\frac{n\eta - \sum_{i=1}^{n} W(d_i)}{\sigma^2}\right) \qquad [4.1]$$

Similarly, the false alarm rate is given by:

$$P_f = \mathbb{P}\left(\frac{1}{n}\sum_{i=1}^{n} N_i^2 > \eta\right) = 1 - \mathbb{P}\left(\sum_{i=1}^{n}\left(\frac{N_i}{\sigma}\right)^2 \leq \frac{n\eta}{\sigma^2}\right) = 1 - \chi_n\left(\frac{n\eta}{\sigma^2}\right) \quad [4.2]$$

4.2.2. *Fusion-based evidential models*

Senouci *et al.* [SEN 14a] consider a more general decision scheme wherein each sensor produces a belief function instead of a binary decision. They do not target a particular sensing model as the model depends on the used sensors and the RoI. Based on [SEN 14a], in this section we detail how belief functions, also called basic belief masses (*bbm*), are modeled and manipulated.

4.2.2.1. *Evidence construction*

When a target is present in the RoI, two states are required to specify whether the target is detected or not: θ_0 (target present and not detected) and θ_1 (target present and detected). Thus, the Frame of Discernment (FoD) is the set $\Theta' = \{\theta_0, \theta_1\}$. To translate the uncertain sensory data into belief functions, we proceed as follows: for a sensor s_i, and relatively to a point $p \in RoI$, we define three inputs: the *bbm* assigned to no detection $m^{\Theta'}_{d_{i/p}}(\theta_0)$, the *bbm* assigned to detection $m^{\Theta'}_{d_{i/p}}(\theta_1)$, and the unassigned *bbm* $m^{\Theta'}_{d_{i/p}}(\theta_0, \theta_1)$. The sum of the three masses always equals one, so there are only two independent masses, we may pick for instance $m^{\Theta'}_{d_{i/p}}(\theta_0)$ and $m^{\Theta'}_{d_{i/p}}(\theta_1)$. The mass $m^{\Theta'}_{d_{i/p}}(\theta_1)$ represents a belief in a target presence at p; the mass $m^{\Theta'}_{d_{i/p}}(\theta_0)$ represents the opposite, and the mass $m^{\Theta'}_{d_{i/p}}(\theta_0, \theta_1)$ reflects the uncertainty of the sensor s_i.

Given the belief $b_{d_{i/p}}$ of a sensor s_i that claims that there is a target located at p, the following mapping is used to acquire the probability masses:

$$m^{\Theta'}_{d_{i/p}}(x) = \begin{cases} (1 - u_i) \cdot (1 - b_{d_{i/p}}) & \text{if } x = \{\theta_0\}, \\ (1 - u_i) \cdot b_{d_{i/p}} & \text{if } x = \{\theta_1\}, \\ u_i & \text{if } x = \{\theta_0, \theta_1\}, \\ 0 & \text{if } x = \{\emptyset\}. \end{cases}$$

where $u_i \in [0, 1]$ is the uncertainty of sensor s_i about its decision.

Alternatively, when a target is not present in the RoI, two states are required to specify whether a target is detected or not: θ_2 (target not present and not detected), θ_3 (target not present and detected). Thus, the FoD is the set $\Theta^{nt} = \{\theta_2, \theta_3\}$. Given the false alarm rate b_{f_i} of a sensor s_i, we translate the uncertain sensory data into belief functions by defining a belief function $m_f^{\Theta^{nt}}$ on the FoD Θ^{nt} as follows:

$$m_{f_i}^{\Theta^{nt}}(x) = \begin{cases} (1-u_i)\cdot(1-b_{f_i}) & \text{if } x = \{\theta_2\}, \\ (1-u_i)\cdot b_{f_i} & \text{if } x = \{\theta_3\}, \\ u_i & \text{if } x = \{\theta_2, \theta_3\}, \\ 0 & \text{if } x = \{\emptyset\}. \end{cases}$$

It is worth pointing out that both formulations $m_{d_{i/p}}^{\Theta^t}$ and $m_{f_i}^{\Theta^{nt}}$ meet Appriou's axioms [APP 91].

4.2.2.2. *Decision combination*

When several decisions are obtained through distinct sensors, a new decision representing the consensus of those independent decisions can be obtained through the combination operation. Let $m_{d_{1\oplus2}}^{\Theta^t}$ be the combination of the two basic belief assignment (*bba*) $m_{d1}^{\Theta^t}$ and $m_{d2}^{\Theta^t}$ produced by the two sensors s_1 and s_2, respectively. Using the Dempster's rule of combination, the *bba* $m_{d_{1\oplus2}}^{\Theta^t} = m_{d1}^{\Theta^t} \oplus m_{d2}^{\Theta^t}$ has the following expression:

$$\begin{aligned}
m_{d_{1\oplus2}}^{\Theta^t}(\theta_0) &= k_{d_{1\oplus2}} \cdot [m_{d_1}^{\Theta^t}(\theta_0)\cdot m_{d_2}^{\Theta^t}(\theta_0) \\
&\quad + m_{d_1}^{\Theta^t}(\theta_0)\cdot m_{d_2}^{\Theta^t}(\theta_0,\theta_1) \\
&\quad + m_{d_1}^{\Theta^t}(\theta_0,\theta_1)\cdot m_{d_2}^{\Theta^t}(\theta_0)] \\
m_{d_{1\oplus2}}^{\Theta^t}(\theta_1) &= k_{d_{1\oplus2}} \cdot [m_{d_1}^{\Theta^t}(\theta_1)\cdot m_{d_2}^{\Theta^t}(\theta_1) \\
&\quad + m_{d_1}^{\Theta^t}(\theta_1)\cdot m_{d_2}^{\Theta^t}(\theta_0,\theta_1) \\
&\quad + m_{d_1}^{\Theta^t}(\theta_0,\theta_1)\cdot m_{d_2}^{\Theta^t}(\theta_1)] \\
m_{d_{1\oplus2}}^{\Theta^t}(\theta_0,\theta_1) &= k_{d_{1\oplus2}} \cdot [m_{d_1}^{\Theta^t}(\theta_0,\theta_1)\cdot m_{d_2}^{\Theta^t}(\theta_0,\theta_1)] \\
m_{d_{1\oplus2}}^{\Theta^t}(\emptyset) &= 0
\end{aligned} \qquad [4.3]$$

where

$$k_{d_{1\oplus 2}} = \frac{1}{1 - (m_{d_1}^{\Theta^t}(\theta_0) \cdot m_{d_2}^{\Theta^t}(\theta_1) + m_{d_1}^{\Theta^t}(\theta_1) \cdot m_{d_2}^{\Theta^t}(\theta_0))}$$

Similarly, $m_{f_{1\oplus 2}}^{\Theta^{nt}} = m_{f_1}^{\Theta^{nt}} \oplus m_{f_2}^{\Theta^{nt}}$ is the combination of two *bbas* $m_{f_1}^{\Theta^{nt}}$ and $m_{f_2}^{\Theta^{nt}}$ produced by two sensors s_1 and s_2, respectively. The *bba* $m_{f_{1\oplus 2}}^{\Theta^{nt}}$ has the following expression:

$$\begin{aligned}
m_{f_{1\oplus 2}}^{\Theta^{nt}}(\theta_2) &= k_{f_{1\oplus 2}} \cdot [m_{f_1}^{\Theta^{nt}}(\theta_2) \cdot m_{f_2}^{\Theta^{nt}}(\theta_2) \\
&+m_{f_1}^{\Theta^{nt}}(\theta_2) \cdot m_{f_2}^{\Theta^{nt}}(\theta_2, \theta_3) \\
&+m_{f_1}^{\Theta^{nt}}(\theta_2, \theta_3) \cdot m_{f_2}^{\Theta^{nt}}(\theta_2)] \\
m_{f_{1\oplus 2}}^{\Theta^{nt}}(\theta_3) &= k_{f_{1\oplus 2}} \cdot [m_{f_1}^{\Theta^{nt}}(\theta_3) \cdot m_{f_2}^{\Theta^{nt}}(\theta_3) \\
&+m_{f_1}^{\Theta^{v}}(\theta_3) \cdot m_{f_2}^{\Theta^{nt}}(\theta_2, \theta_3) \\
&+m_{f_1}^{\Theta^{nt}}(\theta_2, \theta_3) \cdot m_{f_2}^{\Theta^{nt}}(\theta_3)] \\
m_{f_{1\oplus 2}}^{\Theta^{nt}}(\theta_2, \theta_3) &= k_{f_{1\oplus 2}} \cdot [m_{f_1}^{\Theta^{nt}}(\theta_2, \theta_3) \cdot m_{f_2}^{\Theta^{nt}}(\theta_2, \theta_3)] \\
m_{f_{1\oplus 2}}^{\Theta^{nt}}(\emptyset) &= 0
\end{aligned} \qquad [4.4]$$

where,

$$k_{f_{1\oplus 2}} = \frac{1}{1 - (m_{f_1}^{\Theta^{nt}}(\theta_2) \cdot m_{f_2}^{\Theta^{nt}}(\theta_3) + m_{f_1}^{\Theta^{nt}}(\theta_3) \cdot m_{f_2}^{\Theta^{nt}}(\theta_2))}$$

For N sensors, the combination of the N *bbas* $m_{d_1}^{\Theta^t}, \ldots, m_{d_N}^{\Theta^t}$ using the $\oplus$ rule yields a *bba* $m_d^{\Theta^t}$ that has the following expression:

$$\begin{aligned}
m_d^{\Theta^t}(\theta_0) &= 1 - \textstyle\prod_{i=1}^N (1 - m_{d_i}^{\Theta^t}(\theta_0)) \cdot k_d \\
m_d^{\Theta^t}(\theta_1) &= 1 - \textstyle\prod_{i=1}^N (1 - m_{d_i}^{\Theta^t}(\theta_1)) \cdot k_d \\
m_d^{\Theta^t}(\Theta^t) &= \textstyle\prod_{i=1}^N m_{d_i}^{\Theta^t}(\Theta^t) \cdot k_d
\end{aligned} \qquad [4.5]$$

where,

$$k_d = \frac{1}{\prod_{i=1}^N (1 - m_{d_i}^{\Theta^t}(\theta_0)) + \prod_{i=1}^N (1 - m_{d_i}^{\Theta^t}(\theta_1)) - \prod_{i=1}^N m_{d_i}^{\Theta^t}(\Theta^t)}$$

Similarly, the combination of the N *bbas* $m_{f_1}^{\Theta^{nt}}, \ldots, m_{f_N}^{\Theta^{nt}}$ using the $\oplus$ rule yields a *bba* $m_f^{\Theta^{nt}}$ that has the following expression:

$$\begin{aligned} m_f^{\Theta^{nt}}(\theta_2) &= 1 - \prod_{i=1}^N (1 - m_{f_i}^{\Theta^{nt}}(\theta_2)) \cdot k_f \\ m_f^{\Theta^{nt}}(\theta_3) &= 1 - \prod_{i=1}^N (1 - m_{f_i}^{\Theta^{nt}}(\theta_3)) \cdot k_f \\ m_f^{\Theta^{nt}}(\Theta^{nt}) &= \prod_{i=1}^N m_{f_i}^{\Theta^{nt}}(\Theta^{nt}) \cdot k_f \end{aligned} \quad [4.6]$$

where,

$$k_f = \frac{1}{\prod_{i=1}^N (1 - m_{f_i}^{\Theta^{nt}}(\theta_2)) + \prod_{i=1}^N (1 - m_{f_i}^{\Theta^{nt}}(\theta_3)) - \prod_{i=1}^N m_{f_i}^{\Theta^{nt}}(\Theta^{nt})}$$

Using equation [4.6], we plot in Figure 4.1, with 95% confidence intervals, the false alarm rate as a function of the number of sensors (N) involved in the fusion process. Two cases are considered: (i) false alarm rate is a constant value for all sensors (Figure 4.1(a)), and (ii) false alarm rate is generated by a Gaussian distribution $\mathcal{N}(1, 1)$ (Figure 4.1(b)). In both cases, the results obtained show that the false alarm rate of the WSN is reduced to zero by adopting the presented fusion-based evidential model.

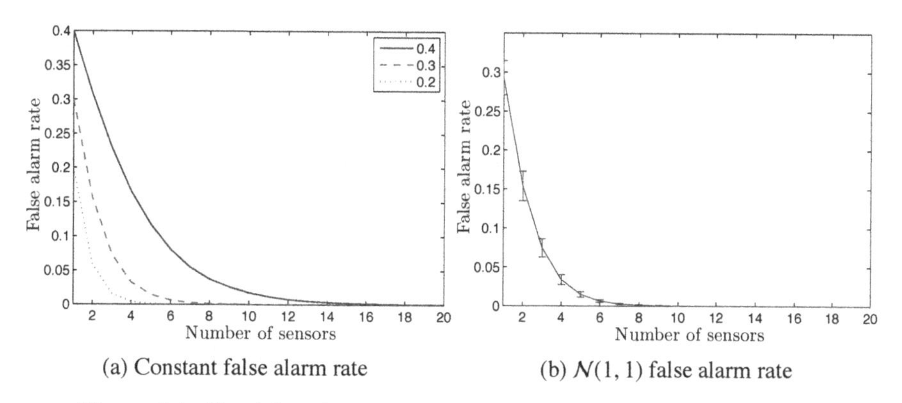

(a) Constant false alarm rate

(b) $\mathcal{N}(1, 1)$ false alarm rate

Figure 4.1. *The false alarm rate vs. the number of sensors. For a color version of the figure, see www.iste.co.uk/senouci/wireless.zip*

4.2.2.3. *Decision-making*

Relatively to a space point $p \in RoI$, two pignistic transformations (denoted by $BetP_{d/p}$ and $BetP_{f/p}$) that permit the construction of the probabilities needed for decision making are constructed.

The pignistic transformation of the *bba* $m_d^{\Theta'}$ denoted by $BetP_{d/p}$ is defined as follows:

$$\begin{aligned} BetP_{d/p}^{\Theta'}(\theta_0) &= m_d^{\Theta'}(\theta_0) + \tfrac{1}{2} m_d^{\Theta'}(\theta_0, \theta_1) \\ BetP_{d/p}^{\Theta'}(\theta_1) &= m_d^{\Theta'}(\theta_1) + \tfrac{1}{2} m_d^{\Theta'}(\theta_0, \theta_1) \end{aligned} \qquad [4.7]$$

Similarly, the pignistic transformation of the *bba* $m_f^{\Theta^{nt}}$, denoted by $BetP_{f/p}$ is defined as follows:

$$\begin{aligned} BetP_{f/p}^{\Theta^{nt}}(\theta_2) &= m_f^{\Theta^{nt}}(\theta_2) + \tfrac{1}{2} m_f^{\Theta^{nt}}(\theta_2, \theta_3) \\ BetP_{f/p}^{\Theta^{nt}}(\theta_3) &= m_f^{\Theta^{nt}}(\theta_3) + \tfrac{1}{2} m_f^{\Theta^{nt}}(\theta_2, \theta_3) \end{aligned} \qquad [4.8]$$

4.3. Fusion-based deterministic deployment problems

Before discussing the different possible formulations of the fusion-based deployment problem, let us revisit how to quantify the quality of coverage provided by a WSN while considering a fusion-based detection coverage model.

4.3.1. *The* (α,β)*-coverage*

For fusion-based probabilistic models, Tan *et al.* [TAN 11] introduced the concept of (α,β)-coverage that quantifies the fraction of the surveillance region where P_d and P_f are bounded by α and β, respectively.

DEFINITION 4.1.– Given two constants, $\alpha \in [0, 1]$ and $\beta \in [0, 1]$, a physical point $p \in RoI$ is (α,β)-covered if the detection probability $P_d(p)$ and the false alarm rate P_f satisfy

$$P_d(p) \geq \alpha \text{ and } P_f \leq \beta \qquad [4.9]$$

where α is the minimum requested detection rate, whereas β is the maximum allowed false alarm rate. α and β are user-specified application-specific parameters. In practice, mission-critical surveillance applications [DUA 04, GU 05, HE 06] require a high detection probability ($\alpha \gg 50\%$) and a low false alarm rate ($\beta < 5\%$). In the case of fusion-based evidential models, the concept of (α,β)-coverage has been redefined by Senouci *et al.* [SEN 14a] as follows:

DEFINITION 4.2.– Given two constants $\alpha \in [0,1]$ and $\beta \in [0,1]$, a physical point $p \in RoI$ is (α,β)-covered if:

$$\begin{aligned} BetP^{\Theta'}_{d/p}(\theta_1) &\geq \alpha \\ BetP^{\Theta''}_{f/p}(\theta_3) &\leq \beta \end{aligned} \qquad [4.10]$$

The (α,β)-coverage of a RoI is defined as the fraction of points in the RoI that are (α,β)-covered.

4.3.2. *Network model and assumptions*

Targets are assumed to appear at a set of known locations referred to as target points defining the set $T \subseteq RoI$. T is determined according to the user requirements. For instance, if full coverage is required, target points can be chosen densely and uniformly in the RoI. The deployment of sensors is constrained to belong to the set $D \subseteq RoI$, consisting of deployment points in the RoI.

For any physical point p, the sensors within a certain distance of p form a cluster and fuse their measurements to detect whether a target is present at p. Thus, for every target point $t \in T$, only a subset of deployed sensors $NS(t) \subseteq D$ participate in the fusion process. NS is a design parameter of the fusion model which is mainly constrained by the communication overhead and will be discussed later.

The fusion-based WSN deployment problem is defined as the problem of covering a set of target points while considering a fusion-based detection coverage model. Of course, other metrics such as network connectivity and lifetime could be considered. In the next sections, we present different formulations of the fusion-based deployment problem.

4.3.3. *Problem 1: handling (α,β)-coverage and cost*

In the simplest form of the fusion-based deployment problem, the number of sensors should be kept to a minimum while satisfying the (α,β)-coverage requirements for all target points also. For each target point $p \in T$, there is an associated predefined minimum event detection probability threshold α_p and a maximum false alarm rate β_p. This problem can be formulated as an optimization problem:

$$\min \sum_{p \in D} x_p \quad [4.11]$$

$$\text{s.t.} \quad BetP^{\Theta^t}_{d/p}(\theta_1) \geq \alpha_p, \ \forall p \in T \quad [4.12]$$

$$BetP^{\Theta^{nt}}_{f/p}(\theta_3) \leq \beta_p, \ \forall p \in T \quad [4.13]$$

$$x_p \in \{0, 1\}, \ \forall p \in D \quad [4.14]$$

Equation [4.11] defines the optimization problem where the objective is to minimize the deployment cost. The solution is constrained in equations [4.12] and [4.13], which requires that each target point is (α,β)-covered. Equation [4.14] defines x_p as a zero-one variable. If $x_p = 1$, then a sensor will be deployed at the deployment point p.

By exploiting the optimal detection thresholds [VAR 96], a similar formulation was introduced in [YUA 08, CHA 11, TAN 11]. The authors assume that targets appear at a set of known physical locations referred to as surveillance spots $T = \{t_j, 1 \leq j \leq K\}$, where $t_j = (x_j, y_j) \in RoI$, is the coordinates of the jth spot. They also assume that only the sensors close to a surveillance spot participate in the data fusion. For any surveillance spot, the fusion region is a disk of radius R centered at the spot. Finally, they define the impact region of a surveillance spot as the disk of radius $2R$ centered at the spot. The authors consider the uniform (α, β)-coverage problem, they set P_{f_p} to its upper bound β, and try to find a sensor placement such that the number of sensors is minimized, subject to $min_{1 \leq j \leq K}\{P_{d_j}\} \geq \alpha$.

All formulations discussed above are non-linear and non-convex optimization problems. Ababnah *et al.* [ABA 10b] treated the deployment problem in the context of optimal control theory. Specifically, they model the deployment problem as a linear quadratic regulator (LQR) problem, with the deployment locations serving as control parameters. Assuming a value fusion scheme and under several approximations, Ababnah *et al.* [ABA 09a] tried to linearize the formulation proposed in [YUA 08] and solve the deployment problem as an optimal control problem. The difference between achieved and requested detection probability over the RoI is modeled as the cost-function in the LQR. It is worth noting that all formulations mentioned above do not discuss the locations of the cluster-heads, they only focus on the sensing quality and ignore network connectivity completely, although sensors are supposed to collaborate. Indeed, these objectives are competing. On the other hand, the (α, β)-coverage objective will desire "spread out" network topology, where sensors are as far apart from each other as possible in order to minimize the sensing overlap. To ensure reliable links and minimize communication cost, sensors must not be located too far from each other. This implies a clustered configuration around cluster-heads with a lot of sensing overlap between sensors, yielding a poor (α, β)-coverage.

4.3.4. *Problem 2: handling (α, β)-coverage, cost and connectivity*

Senouci *et al.* [SEN 14a] focused on generating a good cluster-based topology that consumes less energy while satisfying the (α, β)-coverage and connectivity constraints. We already know that a good clustering algorithm should guarantee that nodes in the same clusters are highly connected while the same are less connected between clusters. There are other issues that need to be addressed when applying clustering approaches, such as the cluster size, the number of orphan (isolated) nodes, etc. Thus, in addition to the sensing quality objective, Senouci *et al.* [SEN 14a] seek to find the best network topology that:

– minimizes the overall number of unnecessary retransmissions;

– minimizes the number of cluster-heads;

– minimizes the number of isolated sensors (ensures that each sensor gets included in a cluster);

– minimizes the number of isolated cluster-heads (ensures that each cluster-head has some cluster members);

– minimizes the overlapping areas among cluster-heads (ensures that the cluster-heads are so distributed or chosen such that there is a minimum overlapping).

On the basis of the connectivity model described previously (see section 1.1.1.2), Senouci *et al.* [SEN 14a] defined the communication cost of a link and a cluster as follows:

DEFINITION 4.3.– The communication cost, of a link ij (denoted by Lc_{ij}) is the expected number of retransmissions needed for a successful communication between the nodes s_i and s_j. Formally:

$$Lc_{ij} = nRT(\|ij\|) = \frac{1}{\mathcal{P}[P_r(\|ij\|) > SS_{min}]}$$

DEFINITION 4.4.– The communication cost of a cluster (denoted by Cc) is the sum of all communication costs of the links connecting cluster members to the cluster-head (denoted by h). Formally:

$$Cc_h = \sum_{s_i \in \text{ cluster members}} Lc_{ih}$$

Formally, the fusion-based deterministic WSN deployment problem is defined as follows:

$$\min \sum_{p \in D} x_p$$

$$\min \sum_{p \in \text{ cluster-heads}} x_p$$

$$\min \sum_{p \in \text{ isolated sensors}} x_p$$

$$\min \sum_{p \in \text{ isolated cluster-heads}} x_p$$

$$\min \sum_{\text{cluster-heads}} \text{cluster-heads overlapping}$$

$$\min \max_{h \in \text{ cluster-heads}} (Cc_h)$$

$$\text{s.t. } BetP^{\Theta^t}_{d/p}(\theta_1) \geq \alpha_p, \qquad \forall p \in T$$

$$BetP^{\Theta^{nt}}_{f/p}(\theta_3) \leq \beta_p, \qquad \forall p \in T$$

$$x_p \in \{0, 1\}, \qquad \forall p \in D$$

This formulation not only enhances the cluster connectivity but also the cluster lifespan, and thus, the network lifespan. It is worth noting that the problem in hand is a multi-objective binary nonlinear and non-convex optimization problem. The next section discusses the sensor placement algorithms proposed in order to solve the different variants of the fusion-based deterministic deployment problem formalized above.

4.4. Sensor placement algorithms

Different placement algorithms were proposed to solve the different formulations of the fusion-based deterministic deployment problem. In this book, we classify the published strategies based on the used fusion-based coverage model. Thus, these strategies can be classified into two groups: (i) fusion-based probabilistic deployment strategies, and (ii) fusion-based evidential deployment strategies. A taxonomy for fusion-based deterministic deployment strategies is depicted in Figure 4.2.

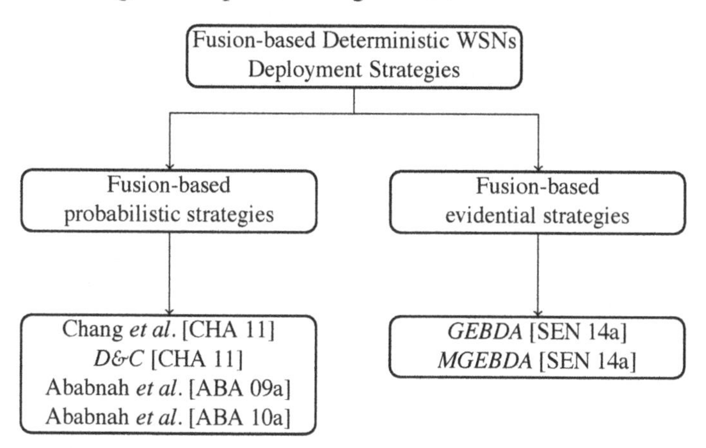

Figure 4.2. *Taxonomy for fusion-based deterministic deployment strategies*

4.4.1. *Fusion-based probabilistic strategies*

Chang *et al.* [CHA 11] proposed a straightforward optimal solution that iterates incrementally N from 1 to search for the optimal sensor placement. In each iteration, $min_{1 \leq j \leq K}\{P_{d_j}\}$ is maximized. Once the constraint $min_{1 \leq j \leq K}\{P_{d_j}\} \geq \beta$ is satisfied, the global optimal solution is found. To solve the nonlinear and non-convex optimization problem in each iteration, Chang *et al.* [CHA 11] apply a non-linear programming solver based on the Constrained Simulated Annealing (CSA) algorithm [WAH 07], which is a global optimal algorithm that converges asymptotically to a constrained global optimum (Theorem 1 in [WAH 07]). However, the complexity of CSA increases exponentially with respect to the number of variables [WAH 07]. Therefore, for a large-scale deployment problem, the global optimal solution becomes prohibitively expensive.

Chang *et al.* [CHA 11] also proposed a relatively low computational cost Divide-and-Conquer (D&C) heuristic. In the divide step, for each surveillance spot t_j, they find the set of spots within the impact region of t_j. In the conquer step, for each surveillance spot t_j, the fewest additional sensors were placed within the fusion region of t_j to cover t_j and the neighboring spots in its impact region. The optimization was implemented by the aforementioned CSA solver.

On the basis of the optimal control theory, Ababnah *et al.* [ABA 10b] proposed a sequential sensor deployment algorithm whose computational complexity is equal to $O(3n^6 + 4n^4)$. In another work, Ababnah *et al.* [ABA 10a] introduced a similar algorithm, with the same computational complexity, while considering the majority or counting rule as a decision fusion rule. Table 4.1 provides a comparative summary of various fusion-based probabilistic approaches discussed above.

4.4.2. *Fusion-based evidential strategies*

Senouci *et al.* [SEN 14a] proposed the Genetic Evidence-Based Deployment Algorithm (GEBDA), an efficient algorithm based on genetic algorithms (GAs) to solve problem 1 (see section 4.3.3). In GEBDA, an individual is a candidate sensor placement that specifies the number and locations of sensors. GEBDA encodes the solutions as strings of bits with length $L = |D|$ from a binary alphabet. Each bit corresponds to a deployment point $p \in D$ and "1" means that a sensor will be deployed at that point. Figure 4.3 shows the binary representation of a placement of 3 sensors within a 4×4 RoI using two mapping algorithms (here $D = RoI$).

Paper	*Computational complexity*	*Fusion model*	*Primary objective*	*Secondary objective*	*Constraint*
[ABA 09a]	$O(3n^6 + 4n^4)$	Value fusion	Max. detection performance	-	Fixed number of sensors
[ABA 10a]	$O(3n^6 + 4n^4)$	Decision fusion	Max. detection performance	-	Fixed number of sensors
[CHA 11]	$O(e^{n^2})$	Value fusion	Max. detection performance	Min. the number of sensors	-
[CHA 11]	$O(n^2me^m)^*$	Value fusion	Max. detection performance	Min. the number of sensors	-

* m is the average number of spots in the impact region.

Table 4.1. *Comparison between various fusion-based probabilistic deployment approaches*

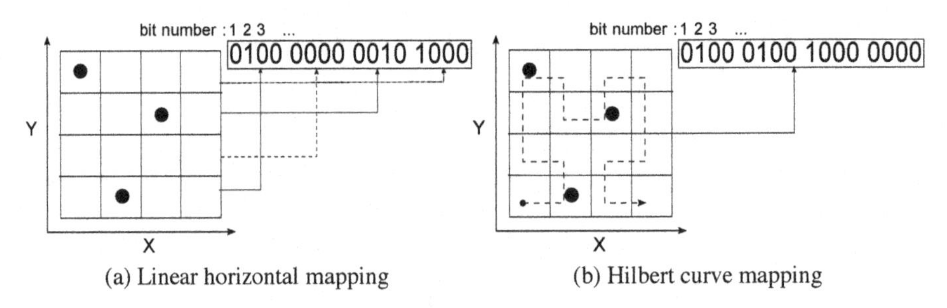

(a) Linear horizontal mapping (b) Hilbert curve mapping

Figure 4.3. *Binary representation of a sensor placement in GEBDA*

GEBDA uses penalties [COE 02] to handle the nonlinear inequality constraints. It assigns a penalty cost for each sensor placement if it does not satisfy the user detection requirements. The infeasibility of an individual represents both the number of active constraints and the extent to which each constraint is violated. A function $viol_d$ that measures the degree of detection, rate constraint dissatisfaction is defined according to equation [4.15].

$$viol_d = \sum_{p \in T} ([\alpha_p - \tilde{\alpha_p}]^+ * \alpha_p). \qquad [4.15]$$

where $\tilde{\alpha_p}$ denotes the generated detection probability at the target point $p \in T$ (computed by equation [4.7]), and $[\alpha_p - \tilde{\alpha_p}]^+$ denotes the projection of $(\alpha_p - \tilde{\alpha_p})$ in $\mathbb{R}^+$. Formally:

$$[\alpha_p - \tilde{\alpha_p}]^+ = (\alpha_p - \tilde{\alpha_p}) \times 1_{\{\alpha_p > \tilde{\alpha_p}\}}.$$

where $1_{\{\alpha_p > \tilde{\alpha_p}\}}$ is the indicator function that is equal to 1 if $\alpha_p > \tilde{\alpha_p}$, otherwise it is equal to 0.

Similarly, equation [4.16] defines a function $viol_f$ that measures the degree of false alarm rate constraint dissatisfaction.

$$viol_f = \sum_{p \in T} \frac{[\tilde{\beta_p} - \beta_p]^+}{\beta_p}. \qquad [4.16]$$

where $\tilde{\beta_p}$ denotes the generated false alarm rate at the target point $p \in T$ (computed using equation [4.8]).

The *viol* function values give an estimate on how far the solution is from the user detection requirements. The weighting schemes introduced in equations [4.15]–[4.16] allow us to give more weight to risky areas.

The fitness function combines the deployment cost, the dissatisfaction of detection, and false alarm constraints. The goal of this fitness function equation [4.17] is to obtain a sensor placement that satisfies the user detection requirements with the minimum number of sensors.

$$fitness = \sum_{p \in D} x_p + C_d \times viol_d + C_f \times viol_f. \qquad [4.17]$$

The executional steps of GEBDA are as follows:

– *Step 1*: create an initial population (of size N) based on random sensor placements;

– *Step 2*: iteratively perform the following sub-steps (called a generation) on the population until a stopping criterion is satisfied:

- analyze the (α, β)-coverage and compute the penalty costs of each sensor placement,

- compute the fitness of each sensor placement: the fitness of each deployment in the population is taken as the weighed sum of the deployment cost plus the penalty costs,

- select sensor placements from the population using the evolution engine,

- create new sensor placements by applying the genetic operators with specified probabilities,

- generate the next population using evolution engine;

– *Step 3*: after the termination criterion is satisfied, the best sensor placement is harvested, and designated as the result of the run.

The stopping criterion can be implemented in several ways, such as:

– the number of generations;

– the amount of time;

– minimum fitness threshold;

– fitness has reached a plateau.

The general structure of GEBDA is shown in Algorithm 4.1.

Algorithm 4.1. GEBDA

Require: RoI, Coverage requirement
Ensure: Optimal placement
1: Set GEBDA parameters
2: Generate random initial population
3: **repeat**
4: Evaluate (α, β)-coverage of each individual in the current population using equation [4.10]
5: Assign fitness value to each individual using equations [4.15]–[4.17]
6: Select sensor placements from the current population using the Evolution Engine
7: Create new sensor placements by applying the genetic operators
8: Generate the next population using the Evolution Engine
9: **until** (Stopping criterion is satisfied)
10: *return* best individual in the current population

GEBDA is based on the Dempster-Shafer theory that captures several characteristics of real-world applications and allows efficient management of the uncertainty related to sensor readings. The effectiveness and efficiency of GEBDA were evaluated using both simulations and experiments. Simulation results reported in [SEN 14a] show that for all small-scale problems, GEBDA reaches the global optimum at very low computation cost. For large-scale problems, GEBDA consistently outperforms many state-of-the-art heuristics.

Using real data traces collected in the DARPA SensIT military vehicle detection experiments [DUA 04], GEBDA has been compared to the fusion-based probabilistic strategy D&C [CHA 11], and the non-fusion probabilistic deployment approach Min_Miss [ZOU 03b]. Obtained results show that, regarding the rate of successful detection, the three strategies ensure the user detection requirements. Furthermore, the Min_Miss strategy tends to employ fewer sensors than the two other strategies as no fusion scheme is being used. However, on average the Min_Miss strategy generates

40.42 % ± 7.76 false alarm rate, which is unacceptable for critical applications that require a low rate of false alarm. The value-fusion scheme used in D&C reduces the false alarm rate, on average, to 23.54 % ± 6.50. GEBDA not only improves sensing coverage by wisely exploiting the collaboration among sensors but also reduces the false alarm rate on average to 5.84 % ± 3.08.

The first conclusion is that fusion improves coverage quality, not coverage. If false alarms are not an issue for the application at hand, probabilistic placement without fusion should be used. Otherwise, a fusion-based placement should be employed. In both cases, the analytical coverage model assumed at the pre-deployment stage should approximate real coverage properties well enough to lead to accurate estimation of application performance. The second conclusion is that evidence-fusion significantly improves coverage quality by exploiting the collaboration among sensors and it outperforms the value-fusion approach.

Problem 2 (see section 4.3.4) is a nonlinear and non-convex optimization problem that is characterized by the presence of many conflicting objectives. Therefore, it is necessary to look at that problem as a multi-objective optimization problem. Using multi-objective optimization techniques, Pareto-optimal (non-dominated or non-inferior) solutions could be obtained for this problem. The solutions belonging to the Pareto-optimal solution set are not dominated by the rest of solutions in the search space. Any solution of the Pareto-optimal front cannot be said to be better than the other solutions in absence of any further information on preference ordering. Therefore, there is a demand to generate Pareto-optimal solutions as much as possible to give more options to the network designer.

On the basis of the multi-objective genetic algorithms NSGA-II [DEB 02], Senouci *et al.* [SEN 14a] devised a new algorithm, the Multi-Objective Genetic Evidence-Based Deployment Algorithm (MGEBDA) to solve problem 2. To encode the problem, MGEBDA uses a similar approach to that of GEBDA. In MGEBDA, an individual is a candidate sensor placement that specifies the number and locations of regular sensors and cluster-heads. The length of this bit string is $L = |D|^2$ as two bits correspond to a deployment point $p \in D$ with the following semantic: "00" no sensor deployed, "01" or "10" a regular sensor deployed, and "11" a cluster-head sensor deployed. Figure 4.4 shows the binary representation of a placement of 3 sensors within a 4×4 RoI using two mapping algorithms (here $D = RoI$).

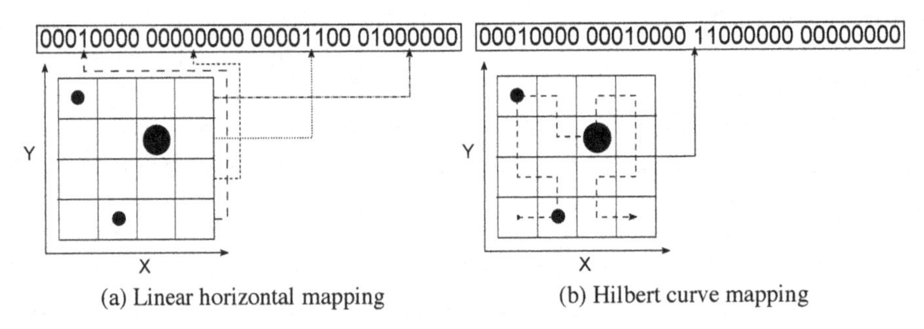

(a) Linear horizontal mapping (b) Hilbert curve mapping

Figure 4.4. *Binary representation of a sensor placement in MGEBDA*

As a constraint handling method, Senouci *et al.* [SEN 14a] use the constrained-domination principle [DEB 02]. The effect of using this constrained-domination principle is that any feasible solution has a better non-domination rank than any infeasible solution. All feasible solutions are ranked according to their non-domination level based on the objective function values. However, among two infeasible solutions, the solution with a smaller constraint violation has a better rank [DEB 02].

The steps involved in the MGEBDA algorithm are described below:

– *Step 1*: initialize the population. The initial population (of size N) is generated using random sensor placements;

– *Step 2*: iteratively perform the following sub-steps (called a generation) on the population until the stopping criterion is satisfied:

- calculate all the objective functions values separately,

- rank the population using the constrained non-dominating criteria,

- calculate the crowding distance of each solution,

- select sensor placements from the population using the evolution engine,

- create new sensor placements by applying the genetic operators with specified probabilities,

- generate the next population using the evolution engine;

– *Step 3*: After the termination criterion is satisfied, the first front of the final population is harvested and designated as the result of the run.

As discussed previously, the fusion radius is a design parameter of the fusion model, which is mainly constrained by the communication overhead. Figure 4.5 plots the deployment cost and the communication cost (as stated by definitions 4.1 and 4.2) required for full (0.9, 0.01)-coverage computed by MGEBDA versus the fusion radius. It should be noted that as MGEBDA generates several Pareto-optimal solutions, we consider, each time, the best solution in terms of deployment cost.

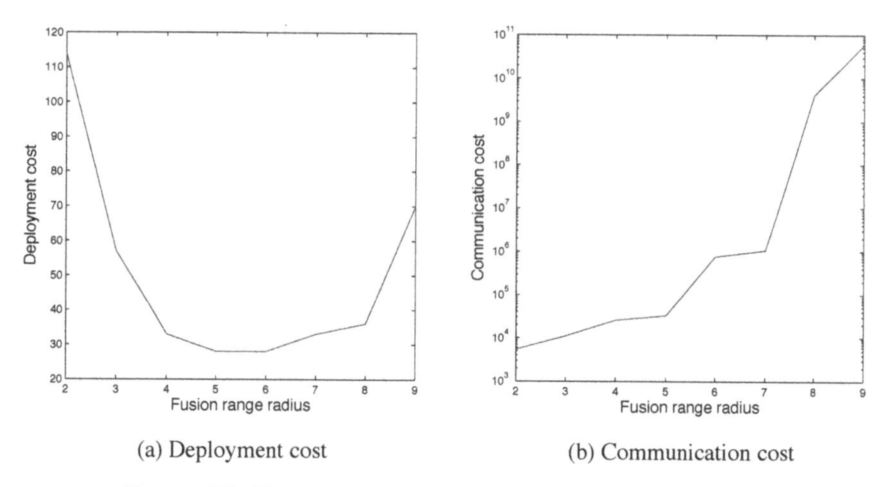

(a) Deployment cost (b) Communication cost

Figure 4.5. *The impact of fusion radius on the performance of MGEBDA*

In Figure 4.5, we can see that: (i) the deployment cost drops rapidly from 114 to 32 when the fusion radius increases from 2 to 6, and gradually increases to 70 when the fusion radius becomes larger, (ii) as the fusion radius increases the communication cost increases exponentially. Intuitively, as the fusion range increases, more distant sensors contribute to the data fusion resulting in better sensing quality, which reduces the deployment cost in terms of regular sensors and cluster-heads. A visualization of a sensor placement generated by MGEBDA is depicted in Figure 4.6. We see clearly that fewer regular sensors and cluster-heads are employed in the case of a fusion radius of 6 as compared to the case of a fusion radius of 2. However, as the fusion range becomes very large, the measurements of sensors far away from the target point contain low-quality information and fusing them leads to lower detection performance, hence the increase in the deployment cost. In all cases, as the distance between sensors and their cluster-head increases, the communication cost increases.

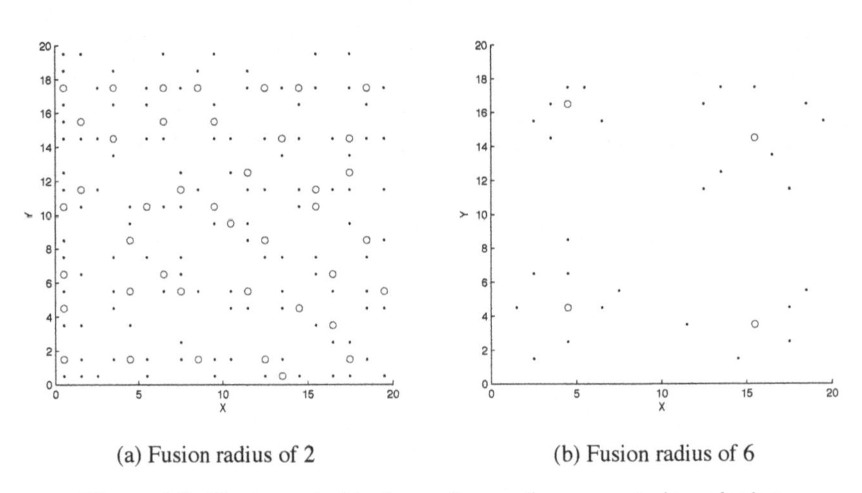

(a) Fusion radius of 2 (b) Fusion radius of 6

Figure 4.6. *The impact of fusion radius on the generated topologies*

4.5. Application: deploying a fusion-based surveillance WSN

In this section, we use the ArduiNet system (see section 3.5) and the fusion-based deployment algorithm MGEBDA [SEN 14a] to deploy a simplified fusion-based indoor surveillance WSN. The ArduiNet system will be deployed as suggested by the output of MGEBDA. We will also report the results obtained.

4.5.1. *Building belief functions*

To build belief functions, we proceed as done previously (see section 3.5.3.1). Figure 4.7 gives an example of a set of mass functions associated to the PIR sensor in an indoor environment. When there is no movement, this sensor provides a value around 500 (value gauged by the manufacturer). Any other measure is equivalent to its symmetrical around 500. Numbers outside the 400 to 600 range denote the detection of a moving object with a high certainty. A projection on the set of mass functions is done in order to obtain the corresponding mass function each time a raw data from that sensor is received. For instance, if the sensor returns a value of 100, then the resulting mass function would have three focal elements: $m(\{\theta_0\}) = 0.05$, $m(\{\theta_1\}) = 0.94$, and $m(\Theta) = 0.01$.

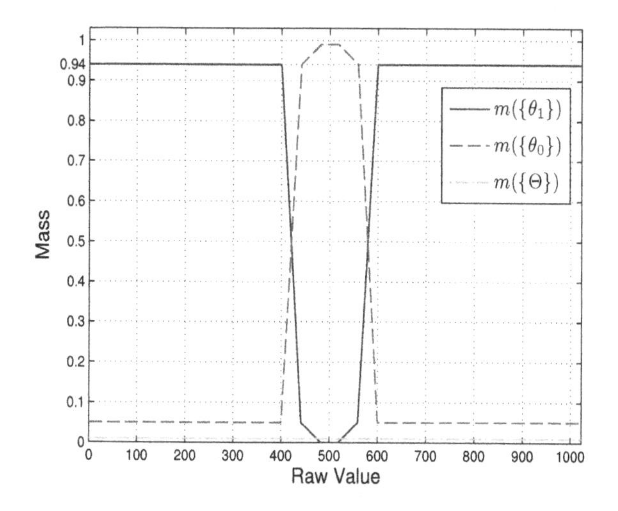

Figure 4.7. *Example of a set of mass functions associated to the PIR Phidgets 1111_0 sensor. For a color version of the figure, see www.iste.co.uk/senouci/wireless.zip*

4.5.2. *Analyzing* (α, β)*-coverage*

We are interested in the performance of ArduiNet as a surveillance system. The main idea of our approach is the determination of the physical and the logical topology of the WSN beforehand to meet the design goals. In other words, we try to estimate, by simulations, the WSN performances before its deployment and we build the best possible topology that meets specific user requirements. The real-world WSN will be deployed as suggested by the obtained topology. For instance, a visualization of a sensor placement generated by MGEBDA for (0.90, 0.01)-coverage is depicted in Figure 4.8.

In individual tests, we deploy ArduiNet as suggested by the output of MGEBDA. We let a single person walk-by at different path in the testbed room, and measure the rate of successful detection and the rate of false alarm.

Each sensor sends its measurements to its cluster-head, which makes the detection decision based on the received measurements. The experiments are repeated more than 5,000 times. The results obtained are summarized in Table 4.2.

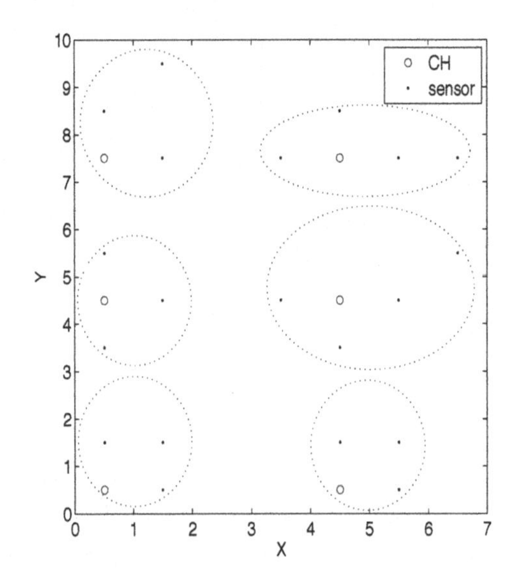

Figure 4.8. *Sensor placement generated by MGEBDA for the lab room scenario*

Requested (α,β)-coverage	*Deployment cost*	*Achieved (α,β)-coverage*
(0.90, 0.01)	26	(0.92, 0.00)
(0.95, 0.01)	29	(0.96, 0.00)

Table 4.2. *Experimental Results*

The results of these tests confirm that MGEBDA is able to meet the desired user detection requirements with the possibility of achieving a zero false positive rate as can be observed in Table 4.2. Obtained results confirm that, by applying a deterministic deployment approach, it is possible to deploy a real-world wireless sensor network with predictable performances.

4.6. Practical issues that need further research

A set of research issues needs to be addressed before surveillance-based WSN applications, such as security surveillance, can become technically feasible and economically practical. This section highlights the major challenges and open problems for future research work.

4.6.1. *Energy efficiency*

Despite the significant advancements in WSNs, energy efficiency remains the most important research challenge. Extremely energy-efficient solutions are required for each aspect of WSN design to deliver the potential advantages of the WSNs technology [AKY 10]. Different solutions at different levels could be adopted to reduce power consumption in various aspects of hardware design, data collecting and processing, network protocols, and operating systems. Examples of popular solutions include compression [RAZ 13], clustering [AFS 14], fusion [NAK 07], topology management [ALS 15], radio duty cycle [BUE 06], and energy harvesting [KHA 15], to name few. In this chapter, we have discussed how exploiting a simple communication cost prediction model could reduce unnecessary transmissions. Precisely, we have showed that it is possible to combine a clustering technique with a fusion model to build the best possible physical topology that enhances the cluster connectivity, reduces energy consumption, and lengthens the network lifetime.

Recent research works advocate that a unified energy-efficient framework that integrates different energy-efficient solutions for different aspects of WSNs should underpin the design of energy-efficient WSNs. In this context, we believe that the deployment strategy should be the main component of such a framework. In other words, the most important optimizations should be done at the design step to build the best energy-efficient topology possible that meets specific user requirements.

4.6.2. *Fusion-based heterogeneous WSN deployment*

Practical surveillance systems may include heterogeneous WSNs that integrate multiple types of sensors running different sensing technologies with different capability/accuracy. As shown in Xing *et al.* [XIN 09], clustering sensors can produce a synergistic effect, allowing sensors with complimentary detection strengths in different scenarios to collaborate. At the design stage, different sensor coverage models should be considered, as different sensing technologies are used for different sensing devices such as cameras, accelerometers, magnets, and PIR sensors. In the literature, several fusion-based deployment approaches assume a data fusion-based modality-specific sensing model [CHA 02, ABA 09a, CHA 11]. These approaches ignore the sensing modalities differences among different sensors, and consequently they do not work in heterogeneous WSNs. Future research in this area should consider generic data fusion sensing models that can function in a wide array of applications. Note that recent research efforts have started to address sensing capability differences among different sensors. For instance, in [KEA 14], the authors focused on capturing sensing capability and clustering the right sensors to meet user accuracy requirements.

4.6.3. *Spatiotemporal coverage in fusion-based WSNs*

Recently, in [AMM 14], Tan and Xing argue that the coverage of WSN has two facets: spatial coverage and temporal coverage. In the context of fusion-based WSNs, spatial coverage is defined as the fraction of area that is (α, β)-covered [TAN 11, SEN 14a], i.e. the fraction of area in which the target can be detected with a detection probability of at least α and a false alarm rate of at most β. The temporal coverage quantifies the timeliness of the network in detecting targets appearing in the RoI [AMM 14]. Although many mission-critical real-time applications require detection delay to be as small as possible, the works discussed previously completely ignore the detection delay.

Tan and Xing [AMM 14] introduced a new concept called α-delay that quantifies the delay of detection under bounded false alarm rate. The authors argue that studying detection delay alone without the consideration of false alarm is meaningless, as a fundamental trade-off between the delay of detection and false alarm rate exists. α-delay was defined as the average number of detection periods before a target is first detected subject to that the false alarm rate of the WSN is no greater than α. Temporal coverage was defined as the reciprocal of α-delay. For instance, if temporal coverage or α-delay approaches one, this means that any target can be detected almost surely in the first detection period after its appearance while the WSN false alarm rate is below α [AMM 14]. Under the assumption that several physical properties of the target signal are known, Tan and Xing [AMM 14] show that data fusion is effective in reducing detection delay and false alarms. Obtained results provide several important guidelines on the design of data fusion algorithms for large-scale WSNs, and could be exploited to design better fusion-based deterministic WSNs deployment strategies.

4.7. Conclusion

This chapter discussed the fusion-based deterministic deployment that is usually employed in the deployment of WSNs for critical applications that impose stringent requirements such as a high detection rate coupled with a low false alarm rate. However, previously discussed deployment approaches compute the optimal sensor placement at the time of the WSN setup and do not consider dynamic changes during the WSN operation. For instance, user requirements can vary over time. One possible approach to deal with such dynamic variations is to dynamically reposition the sensors while the network is operational. This approach, known as dynamic deployment, is the focus of the next chapter.

5

Dynamic Deployment

This chapter considers the deployment of mobile wireless sensor networks. Two research issues are discussed: movement-assisted sensor deployment (i.e. self-deployment) and sensor relocation, both of which involve autonomous sensor movement. Recent literature pertaining to the WSN self-deployment and sensor relocation problems is reviewed. Different proposed algorithms are categorized, and a classification of the most recent deployment techniques is provided. Following that, the high-level classification is detailed, the corresponding techniques are described, summarized and compared. Finally, practical issues that need further research are discussed.

5.1. Why dynamic deployment?

The emerging hardware techniques have promoted the development of mobile sensors [LAI 05]. Indeed, with the advances in mobile devices, some of these sensors, such as Robomote [SIB 02], are able to move on their own. Sensors could also be attached to a moving entity (e.g. Figure 1.7). In Chapters 3 and 4, we have discussed static deterministic deployment approaches that strive to deterministically place the sensors in order to meet the desired performance goals. All of the deterministic deployment approaches compute the optimal sensor placement initially and do not consider moving sensors once they have been deployed. These approaches are static in the sense that the optimal sensor placement is computed at the time of the WSN setup and does not consider dynamic changes during the WSN operation. For instance, user requirements can vary over time, and the available network resources may change as new sensors join the network, or as older ones run out of energy. These situations show that static deterministic deployment approaches are not the recommended choice for many WSN applications.

Consequently, dynamically repositioning the sensors while the network is operational is necessary to deal with dynamic variations in user interests, network resources, and the surrounding environment. For instance, when many sensors in the vicinity of the sink run out of batteries, redundant sensors from other parts of the RoI can be relocated to replace the dead sensors in order to improve the performance of the network. In the case of target tracking applications, sensors can be relocated close to the target to increase the fidelity of the sensor's data. Other examples include recovering from connectivity problems, extending network lifespan, reducing target detection delay, to name but a few.

With the ability to move independently, mobile WSNs are able to self-deploy and self-repair. Several research issues arise such as movement-assisted sensor deployment and sensor relocation. It is worth pointing out that movement-assisted sensor deployment and sensor relocation are two different research issues, both of which involve autonomous sensor movement. They share, in most cases, a common goal, that is, to improve the overall network performance. In movement-assisted sensor deployment, mobile sensors are able to self-configure after early deployment so that a better arrangement is achieved and the network performances are increased. On the other hand, in sensor relocation, the repositioning of nodes dynamically, while the network is operational, is needed to improve the network performance. For instance, the loss of several sensors in the vicinity of the sink due to exhaustion of their batteries can break communication paths in the network and make some sensors unreachable. In the worst case, the network may become partitioned into multiple sub-networks and become dysfunctional. In this case, some redundant sensors from other parts of the RoI can be identified and relocated to replace the dead sensors in order to improve the network lifetime. The network recovery should be both quick and lightweight.

5.2. Movement-assisted sensor deployment algorithms

Although the movement-assisted sensor deployment problem and its derivative coverage problem have been interpreted in a variety of ways in existing literature, a general interpretation is defined as follows: given a mobile WSN randomly deployed in the RoI, the problem is *where to move and how to move mobile sensors efficiently so that the final deployment meets the design goals of the network (coverage, connectivity, fault-tolerance, etc.).*

As a fundamental issue in WSNs, movement-assisted sensor deployment is a research topic that has attracted much attention in recent years [GHO 06, YOU 08, AZI 09, WAN 09, CHE 09, SIL 14, SEN 15b]. Designing a scheme to relocate the mobile sensors depends strongly on the desired performance goals. Most research into nodes mobility focuses on designing algorithms to relocate sensors in order to enhance the quality of monitoring. The proposed algorithms, mainly, strive to spread sensors in the RoI to maximize the covered area. Sensor mobility is exploited essentially to obtain a new stationary configuration, after the displacement of sensors to their desired locations, which improves coverage. The main difference among these schemes is *how exactly the sensors' new positions are computed.*

In this chapter, we explore the published movement-assisted deployment strategies. Based on the relocation scheme, we classify the identified strategies into seven classes, namely: 1) virtual forces-based approach, 2) pattern-based approach, 3) grid-quorum based approach, 4) computational geometry-based approach, 5) fuzzy logic-based approach, 6) metaheuristic-based approach, and 7) bio-inspired approach. An eighth class, miscellaneous approaches, includes the strategies that do not share their principle with the aforementioned seven classes. A taxonomy for movement-assisted sensor deployment algorithms is depicted in Figure 5.1.

Before discussing the different movement-assisted deployment strategies, we enumerate the set of metrics that can be used to compare the different strategies. We later use such metrics to compare the surveyed self-deployment algorithms. We have identified the following metrics:

– sensor coverage model (CM): can either be Binary (B) or Probabilistic (P);

– coverage degree (CD): represents the degree of coverage that the algorithm can ensure; it can be controlled (T) or singular (S);

– distributed *vs*. centralized (D/C): defines if the algorithm is executed by each node of the network or only by specific nodes;

– R_c *vs*. R_s: defines the relationship between the sensing range and the communication range (NDR: no direct relation);

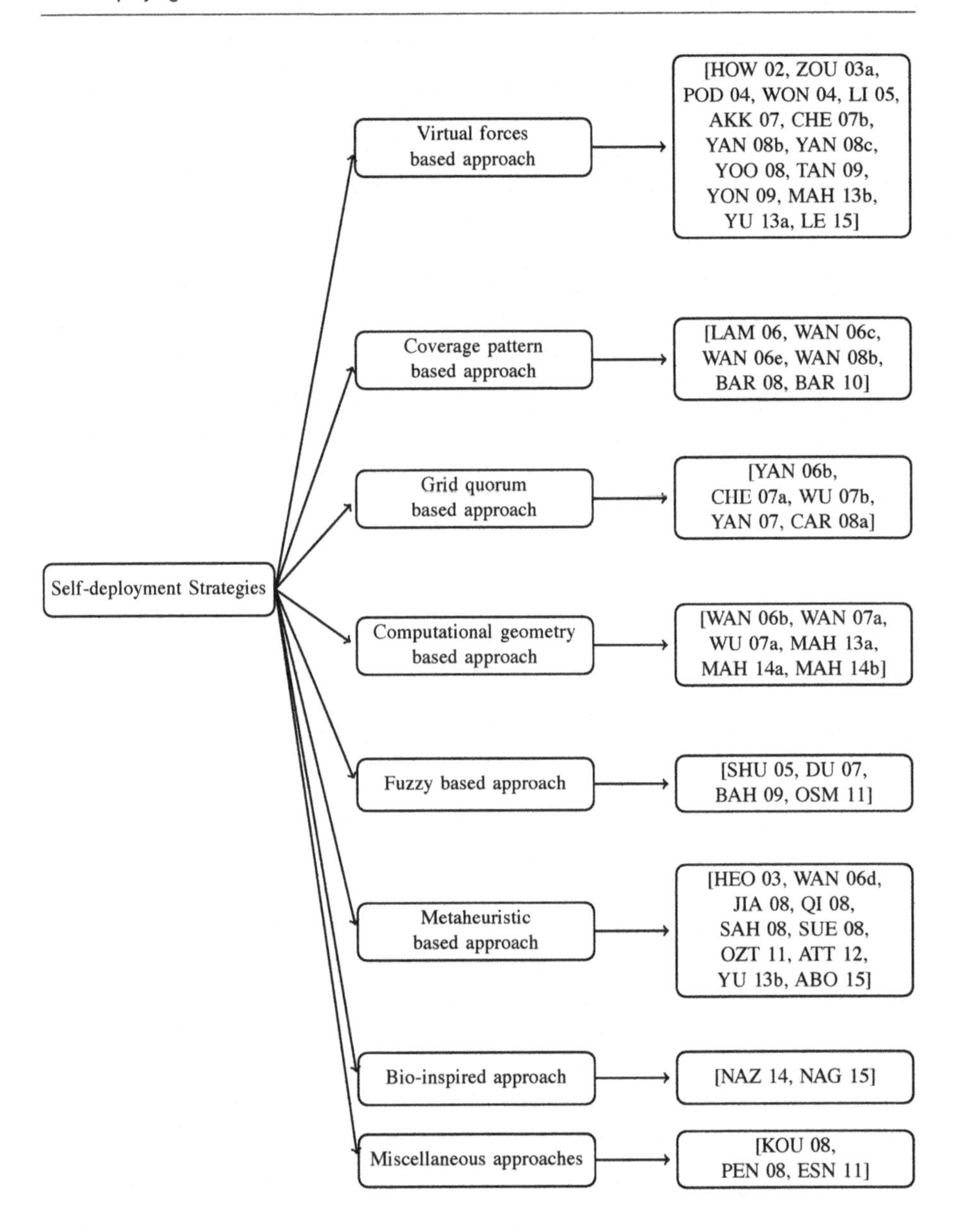

Figure 5.1. *Taxonomy for self-deployment strategies*

– primary objective: defines the main objective that the algorithm tries to reach;

– termination condition: defines the convergence criteria of the algorithm. It can be:

- the coverage degree: the algorithm initially fixes a coverage degree to reach. At each iteration, the algorithm compares the actual coverage degree with the prefixed one. This condition is usually used in centralized algorithms,

- the maximum number of iterations: the algorithm is executed for a prefixed number of iterations,

- stability: in distributed algorithms, when the movement distance of a sensor becomes lower than a threshold distance for a predetermined number of iterations, it involves that this node is in a stability state and that it has to stop moving,

- oscillation: if the node makes a predetermined number of oscillations around the same position, the algorithm must terminate,

- balanced load: this condition is used in the algorithms that belong to the grid quorum approach. The algorithm terminates when the number of nodes in each cluster of the grid is balanced,

- local hole coverage: each node locally estimates the size of its coverage hole and tries to fix it;

– efficiency and effectiveness metrics: complexity, communication overhead, convergence speed, and the number of moves of all mobile sensors.

5.2.1. *Virtual force based approach*

In the virtual force-based node self-deployment approach [HOW 02, ZOU 03a, POD 04, WON 04, LI 05, AKK 07, CHE 07b, YAN 08b, YAN 08c, YOO 08, TAN 09, YON 09, MAH 13b, YU 13a, LE 15], sensors behave as electromagnetic particles. A virtual force concept is adopted in order to relocate sensors to their next position. The inter-sensor force can be either attractive or repulsive. The nature of the force depends essentially on the Euclidean distance between each pair of sensors, where each sensor is supposed to know its position and the list of all its neighbors. If two sensors are too close to each other, where the closeness is measured by a predetermined threshold, they exert repulsive forces on each other. This ensures the minimization of the overlapping coverage area between the two sensors. On the other hand, if a pair of sensors are too far apart from each other (according to the pre-determined threshold distance), they exert an attractive force on each other, which will ensure a global uniform sensor placement. The final movement of a sensor is dictated by the combined force at that sensor due to sensors in its neighborhood. In Figure 5.2, the overall force $\overrightarrow{F_1}$ on s_1 is given by $\overrightarrow{F_1} = \overrightarrow{f_{12}} + \overrightarrow{f_{13}}$. Where d_{th} is the threshold distance, $||s_1 s_2|| > d_{th}$, $||s_1 s_3|| < d_{th}$, and $||s_1 s_4|| = d_{th}$.

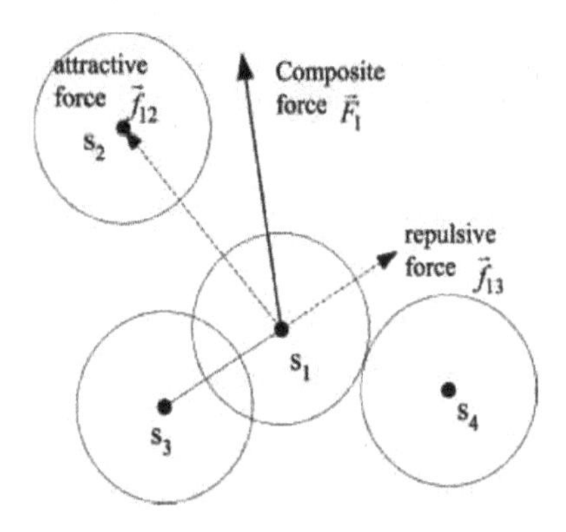

Figure 5.2. *Example of virtual forces*

In the literature, we find a significant number of algorithms that exploit the virtual forces principle for the relocation of mobile sensors. Zou and Chakrabarty [ZOU 03a] proposed a cluster model where each sensor is exposed to three sources of force, which are its neighbors, obstacles and holes in the RoI. Yang *et al.* [YAN 08b] added in their force model a new source of force which is a virtual point situated inside the RoI named the Deployment Center (DC). DC exerts an attracting force on all sensors. Sensors receiving forces from DC move toward it. Wong *et al.* [WON 04] adopt a concept of back off delay in order to solve the problem of simultaneous movement of sensors. This model reduces the collision between the sensors, but it suffers from excessive energy consumption. Yong and Li [YON 09] proposed a distributed algorithm based on the definition of the local and global density around a sensor to calculate the total force exerted by all its neighbors.

Poduri and Sukhatme [POD 04] assumed that every sensor has at least k neighbors in order to ensure the connectivity of the whole network after the sensors' movement. Howard *et al.* [HOW 02] exploited a model based on the potential field theory where each sensor attempts to find a balance state that is the result of the forces exerted by its neighborhood. Akkaya and Younis [AKK 07] used the sensors' mobility in order to guarantee the overall network connectivity. Li *et al.* [LI 05] exploited the sensors' relocation for the purpose of target tracking where they attempted to find the ideal combination of forces that ensure the overall coverage of the network. Mahfoudh *et al.* [MAH 13b] consider both coverage and connectivity in the presence of obstacles. Yu *et al.* [YU 13a] defined the relationship of adjacency of nodes by Delaunay triangulation, and devised a new sensor deployment algorithm based on van der Waals force. Li *et al.* [LE 15] investigated the coverage of a moving phenomenon, and proposed a virtual force-based scheme, namely VirFID. To maximize the weighted sensing coverage, sensors moved toward the positions where more interesting sensing data could be obtained by utilizing the virtual force, which is calculated based on the distance between sensors and sensed values in the RoI.

We performed a qualitative comparison between the different algorithms discussed earlier. Table 5.1 provides a comparative summary of the characteristics of the virtual forces based algorithms.

5.2.2. *Computational geometry based approach*

In this approach [WAN 06b, WAN 07a, WU 07a, MAH 13a, MAH 14b, MAH 14a], a geometric computation was used in order to discover the existence of coverage holes in the RoI and settle the target position where the mobile sensor will be relocated from densely deployed areas to the sparsely ones. The most commonly used computational geometry approaches are the Voronoi diagram and Delaunay triangulation.

The Voronoi diagram [FOR 97] is an important data structure in computational geometry. It represents the proximity information about a set of geometric nodes. The Voronoi diagram of a collection of nodes, partitions the space into polygons. Every point in a given polygon is closer to the node in this polygon than to any other node (Figure 5.3). Based on the sensors' positions, the Voronoi diagram of the network is constructed and the decision whether the sensors need to reposition or to stay will be made based on the diagram.

Algorithm	Coverage model	Coverage degree	D/C	R_c *vs.* R_s	Primary objective	Termination condition	Complexity	Communication overhead	Convergence speed	Sensor movement
VFA [ZOU 03a]	B & P	T	C	NDR	Maximize the overall coverage	Coverage degree	Medium	High	Fast	High
EVFA [CHE 07b]	B	T	C	NDR	Maximize the overall coverage	Coverage degree	Medium	High	Fast	Medium
IVFA [CHE 07b]	B	T	C	NDR	Maximize the overall coverage	Coverage degree	Medium	High	Fast	Medium
TIVFA [LI 05]	P	T	C	NDR	Target tracking	Maximum number of iterations	Medium	High	Fast	Medium
ESD [YAN 08b]	B	T	D	NDR	Eliminate holes and partitions	Stability state	Medium	Low	Fast	High
BODVFA [WON 04]	B	T	D	$R_c \gg R_s$	Maximize the overall coverage	Maximum number of iterations	Low	Low	Medium	Medium
DSSA [YON 09]	B	T	D	$R_c = 2R_s$	Maximize the lifetime of the network	Stability or oscillation	Medium	Low	Medium	High
C^2AP [AKK 07]	B	S	D	NDR	Maximize the overall coverage while maintaining connectivity	Stability state	Medium	Low	Medium	High
EVFA [YOO 08]	P	T	C	$R_c > 2R_s$	Extend the lifetime of the network	Coverage degree	Medium	High	Fast	Medium
VFDEA [YAN 08c]	B	T	D	NDR	Eliminate holes and maximize connectivity	Maximum number of iterations	Low	Medium	Medium	High
Howard *et al.* [HOW 02]	B	S	D	NDR	Improve the overall coverage while minimizing the convergence time	Stability state	Medium	Medium	Fast	High
DVFA [MAH 13b]	B	S	D	$R_c \geq 2R_s$	Ensure coverage and connectivity	Stability state	Medium	High	Medium	High
VirFID [LE 15]	B	S	D	NDR	Maximize the weighted coverage	Stability state	Medium	High	High	High

Table 5.1. *Comparison between the virtual forces based algorithms*

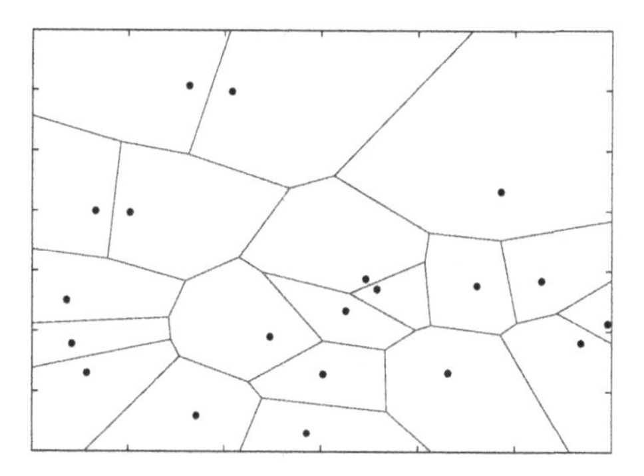

Figure 5.3. *Example of Voronoi diagram*

Wang *et al.* [WAN 06b] proposed three relocation protocols based on the use of the Voronoi diagram, namely: *VEC*, *VOR* and *MiniMax*. The three algorithms attempt to reach a uniform distribution of sensors in the RoI by the relocation of the sensors from dense regions to sparse ones relying on the geometric properties of the Voronoi diagram. Using the same principle, Mahboubi *et al.* [MAH 14b] proposed edge-based and vertex-based strategies wherein the sensors find coverage holes within their Voronoi polygons and then move in the appropriate direction to minimize them. In [WAN 07a], Wang *et al.* presented a protocol called the Bidding protocol. The authors supposed a hybrid network composed of static and mobile sensors. The static sensors identified the coverage holes by means of the Voronoi diagram then the mobile sensors relocated to fill the voids identified by the static ones. In [LIA 15], Liao *et al.* used the Voronoi diagrams of targets to deploy mobile sensors with minimum movement to form a WSN that provides both target coverage and network connectivity.

Mahboubi *et al.* [MAH 13a] considered the case of non-uniform coverage with non-identical sensors. Based on the multiplicatively weighted Voronoi (MW-Voronoi) diagram, they developed three algorithms: *MWV*, *MWP* and *MDW*. According to the proposed algorithms, each sensor moves iteratively in a way such that the prioritized uncovered area in its MW-Voronoi region is reduced. It is worth mentioning that for the case when all sensors have the same sensing capability and the weight of every point in the RoI is the same, the *MDW* strategy will be the same as the *VOR* strategy proposed in [WAN 06b]. Using the MW-Voronoi diagram, Mahboubi *et al.* [MAH 14a] proposed three other relocation strategies: *WVB*, *Minmax-curve* and *Maxmin-curve*. These strategies find appropriate locations for the sensors to increase sensing coverage. These locations are determined such that they are neither too close to each other and the boundaries of their MW-Voronoi regions, nor too far from them [MAH 14a].

Table 5.2 provides a comparative summary of the characteristics of the computational geometry based algorithms.

5.2.3. *Fuzzy based approach*

The fuzzy based approach [SHU 05, DU 07, BAH 09, OSM 11] proposed using a fuzzy logic system to control the sensor movement. It parts from the conviction that fuzzy reasoning is a powerful tool that has the capability to handle uncertainty and ambiguity; it can present more smooth results than common proposed methods in reasoning. The fuzzy system fixes several rules based on the definition of a selection of antecedents that can be, for example, the number of neighbors or the Euclidean distance between a pair of sensors. The output provided by the system is the direction and the next-step move distance for each sensor.

Shu *et al.* [SHU 05] presented a relocation algorithm based on a fuzzy system. In order to identify the movement distance for each sensor, the system uses two antecedents: the Euclidean distance and the number of neighbors. To determine the movement direction the authors appeal to the Coulomb's law. Bahareh *et al.* [BAH 09] used the number of nodes in the neighborhood of a node and the distance from the closest obstacle to settle the system output, which is the priority order of movements for all the sensors. Du *et al.* [DU 07] proposed a fuzzy logic based distributed decision-making algorithm. The mobile sensor uses the fuzzy logic algorithm to determine the best location that it should move to. Their utility function includes three factors: the movement cost, the coverage gain, and the connectivity gain. Osmani [OSM 11] proposes two redeployment algorithms, namely: *FReD* and *FSPNS*, both based on a fuzzy system. In addition to the density of neighbors and average distance from neighbors, Osmani [OSM 11] considers a third antecedent which is the COVERAGE-FACTOR defined as the intersection of the Voronoi polygon of the sensor and its sensing disk. The system output is the movement policy. Table 5.3 provides a comparative summary of the characteristics of the fuzzy based algorithms.

Algorithm	Coverage model	Coverage degree	D/C	R_c *vs.* R_s	Primary objective	Termination condition	Complexity	Communication overhead	Convergence speed	Sensor movement
Vec, Vor, and MiniMax [WAN 06b]	B	S	D	$R_c \geq 2R_s$	Heal the coverage holes	Local coverage of the hole	Medium	Medium	Medium	High
Wang *et al.* [WAN 07a]	B	S	D	$R_c \geq 2R_s$	Heal the holes created by the static nodes	Local coverage of the hole	Medium	High	Fast	Medium
MWV, MWP, and MDW [MAH 13a]	B	S	D	$R_c \geq 2R_s$	Heal the prioritized uncovered area	Local coverage of the hole	High	Medium	Medium	High
Maxmin-vertex, Maxmin-edge, Minimax-edge, and VEDGE [MAH 14b]	B	S	D	NDR	Heal the coverage holes	Local coverage of the hole	Medium	Medium	Medium	High
WVB, Minmax-curve, and Maxmin-curve [MAH 14a]	B	S	D	NDR	Improve the overall coverage	Local coverage of the MW-Voronoi region	High	Medium	Medium	High
TV-Greedy [LIA 15]	B	T	C	NDR	Provide targets coverage and network connectivity	Coverage degree	Medium	Medium	Fast	Medium

Table 5.2. *Comparison between the computational geometry based algorithms*

Algorithm	Coverage model	Coverage degree	D/C	R_c *vs.* R_s	Primary objective	Termination condition	Complexity	Communication overhead	Convergence speed	Sensor movement
FOA [SHU 05]	B	S	D	NDR	Maximize the overall coverage	Stability state	Medium	Medium	Medium	High
FBPC [BAH 09]	B	S	D	$R_c \gg R_s$	Maximize the overall coverage	Stability state	Medium	Medium	Medium	High
Du *et al.* [DU 07]	B	S	D	NDR	Enhance coverage and connectivity	Maximum number of iterations	Low	Low	Fast	Medium
FReD and FSPNS [OSM 11]	B	S	D	NDR	Maximize the overall coverage	Stability state	Medium	Medium	Medium	High

Table 5.3. *Comparison between the fuzzy based algorithms*

5.2.4. *Metaheuristic based approach*

Algorithms belonging to this class [HEO 03, WAN 06d, JIA 08, QI 08, SAH 08, SUE 08, OZT 11, ATT 12, YU 13b, ABO 15] use the power of metaheuristics in order to settle the position, the direction, and the movement speed of a mobile sensor. Ant Colony (AC), Genetic Algorithms (GA), Particle Swarm Optimization (PSO), and Simulated Annealing (SA) are examples of such algorithms.

Genetic Algorithms (GA) are very robust optimization techniques used to find a global solution for a complex problem. Contrary to the other relocation approaches, the GA based approach does not give a unique solution to the problem. It first generates a population of candidate solutions to the problem. Each individual of the population is characterized by its genome, which is composed of one or several chromosomes represented by the speed, position and movement destination of the sensor. The genome is used in a fitness function that evaluates the candidate solutions to settle the best solution for the current iteration. Jiang *et al.* [JIA 08] propose a distributed algorithm that uses the node position and the positions of its neighbors to represent the sensor in the fitness function. The algorithm supposes a dense initial deployment to ensure the coverage of the whole deployment area. In [SUE 08], the author proposes a clustering based algorithm. It is executed in five steps: initialization, collection, recombination, mutation, and termination. Sahin *et al.* [SAH 08] present a genetic algorithm based on the use of virtual forces. The node's genome is defined as its speed and movement direction that are calculated by means of virtual inter-node forces. Attea *et al.* [ATT 12] exploited the multi-objective evolutionary algorithm based on decomposition (MOEA/D) to provide a trade-off between the target's coverage and the nodes' traveled distance.

In addition to GA, various metaheuristics have been exploited by researchers for WSN self-deployment. In [QI 08], the ant colony theory is used to find the optimization scheme for path planning and deployment of mobile sensors. Wang *et al.* [WAN 06d] proposed a new approach based on parallel particle swarm optimization (PPSO). The SA approach has been considered in [HEO 03]. In [OZT 11], the Artificial Bee Colony (ABC) algorithm, a swarm based intelligent method inspired by modeling the foraging behavior of honey bees, is used for the self-deployment of WSNs. Simulation results show that the proposed algorithm obtains better WSN deployments than the PSO algorithm. A modified ABC algorithm introducing forgetting and a neighbor factor in the onlooker bee phase and backward learning in the scout bee phase is proposed in [YU 13b]. Simulation results presented in [YU 13b] show that the proposed approach has higher coverage rate and less energy consumption than the original one. A centralized self-deployment algorithm that utilizes the Multi-objective Immune Algorithm (MIA) has been proposed in [ABO 15]. Simulation results presented by the authors show that the proposed algorithm outperforms the PSO-based one.

Table 5.4 provides a comparative summary of the characteristics of the metaheuristic based algorithms.

5.2.5. *Coverage pattern based approach*

According to the coverage pattern based approach [LAM 06, WAN 06c, WAN 06e, WAN 08b, BAR 08, BAR 10], the final positions where the mobile sensors will be relocated are pre-calculated based on a pre-selected coverage pattern which can ensure the overall coverage and connectivity of the whole network. The RoI is partitioned into a virtual grid in the form of the selected coverage pattern. There are three types of grids commonly used: triangular lattice (Figure 5.4(a)) square grid (Figure 5.4(b)) and hexagonal grid (Figure 5.4(c)).

Triangular lattice is the best among the three kinds of grids as it has the smallest overlapping area. Square grid provides a fairly good performance for any parameters, while hexagonal grid is the worst of all as it has the biggest overlapping area. In addition to the type of grid, the size of grid also plays an important role. The size of grid has to be chosen based on how dense the WSN is going to be. For a highly dense network, a small size grid helps in reducing coverage holes thus providing a better result. However, in a sparse network, a large grid-size is better as it will avoid overlapping of sensors' sensing range, therefore ensuring full utilization of their sensing capabilities. The vertices of the final grid are the target positions where the mobile sensors will be relocated. This approach presents the advantage that as is based on the sensing range of each sensor we can compute the exact number of sensors to deploy initially which can ensure the overall coverage of the RoI.

Algorithm	Coverage model	Coverage degree	D/C	R_c *vs.* R_s	Primary objective	Termination condition	Complexity	Communication overhead	Convergence speed	Sensor movement
Yu *et al.* [YU 13b]	B	S	D	NDR	Maximize the overall coverage	Maximum number of iterations	High	High	Medium	High
LODICO [JIA 08]	B	S	D	$R_c \geq 3R_s$	Maximize the overall coverage	Stability state	High	Medium	Medium	Medium
FGA [SAH 08]	B	S	D	NDR	Ensure the connectivity of the network	Stability state	High	Medium	Slow	High
Yulai [SUE 08]	B	S	C	NDR	Maximize the overall coverage	Maximum number of iterations	High	High	Medium	High
MOEA/D [ATT 12]	B	S	C	NDR	Provide targets coverage	Maximum number of iterations	High	Low	Medium	Medium
Abo-Zahhad *et al.* [ABO 15]	B	S	C	NDR	Maximize the overall coverage	Stability state	High	Low	Medium	Medium

Table 5.4. *Comparison between the metaheuristic based algorithms*

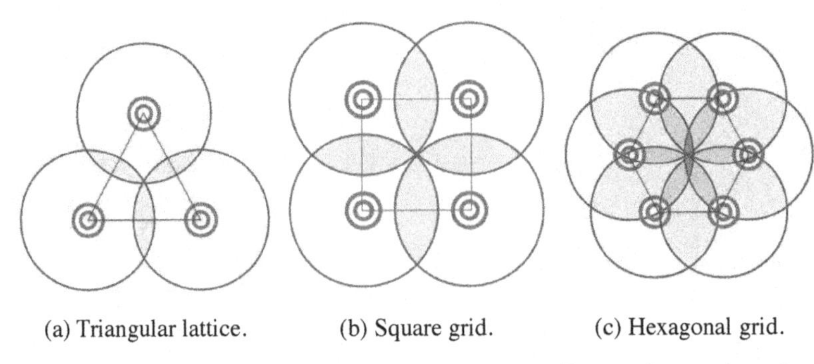

(a) Triangular lattice. (b) Square grid. (c) Hexagonal grid.

Figure 5.4. *Commonly used coverage patterns. For a color version of the figure, see www.iste.co.uk/senouci/wireless.zip*

Wang *et al.* [WAN 06e] proposed a centralized algorithm based on a lozenge pattern. The algorithm supposed that the minimum number of deployed nodes must be greater than $\frac{2A}{3\sqrt{3}r^2}$ where A is the area of the RoI, and r is the sensing range of a node. Lam and Liu [LAM 06] presented a distributed algorithm named *Isogrid* that used a triangular pattern for a uniform coverage of the RoI. Wang *et al.* [WAN 06c] assumed a hexagonal pattern where each node is charged to relocate six of its neighbors in order to form a perfect hexagon of coverage. Bartolini *et al.* [BAR 08] proposed an algorithm named *Snap & Spread* based on the use of a hexagonal pattern. The algorithm attempts to construct a uniform mesh through the definition of the Snap and Spread activities. In a more recent work [BAR 10], the same authors also presented the *Push & Pull* (*P&P*) algorithm, which is an improvement of the *Snap & Spread* algorithm, and uses the same hexagonal pattern. Wang *et al.* [WAN 08b] transformed the problem of node relocation to a matching problem on the basis of regular pattern which can ensure the connectivity of the network.

Table 5.5 provides a comparative summary of the characteristics of the coverage pattern based algorithms.

5.2.6. *Grid quorum based approach*

In the grid quorum based approach [WAN 05a, YAN 06b, CHE 07a, WU 07b, YAN 07, CAR 08a], the sensor deployment problem was brought to a load balance problem as in traditional parallel processing, where each region corresponds to a processor, and the number of sensors in a region corresponds to the load. The RoI is partitioned into a two-dimensional grid of cells through clustering. Each cluster corresponds to a square region and has a cluster-head that is in charge of bookkeeping and communication with adjacent cluster-heads. In contrast to the pattern-based and the virtual forces based schemes, the grid quorum-based approach does not try to find a final target position where the mobile node must be relocated, instead it attempts to define a target cluster in the grid where the node can be localized anywhere. A scan operation is then used to calculate the average load (number of sensors in a cluster) and then to determine the amount of overload and under-load in clusters. Load is shifted from overloaded clusters to underloaded clusters under the constraint of minimizing the overall traversed distance.

Yang *et al.* [YAN 06b] transformed the load balance problem to a matching problem in a bipartite graph. They proposed a centralized solution named the Hungarian method which they compared with a hybrid approach called *SMART* [YAN 07]. *SMART* uses a scan technique that is performed on two steps, first on rows and then on columns, to settle the direction of the node movement. Chellappan *et al.* [CHE 07a] rejected the fact that a sensor can move an unlimited distance to reach its final position. They proposed a model where each sensor can only make a flip movement to a limited distance. In [CAR 08a], Cardei *et al.* presented a distributed algorithm that used a sector scan. The final deployment obtained by the algorithm is not uniform. The density of sensors around the sink is greater than the density in the other regions.

Table 5.6 provides a comparative summary of the characteristics of the grid quorum based algorithms.

5.2.7. *Bio-inspired approach*

Solutions drawn from nature can be useful to address complex problems with simple yet effective approaches [DRE 10]. In the bio-inspired approach [NAZ 14, NAG 15] the main idea is to exploit solutions drawn from nature to deploy WSNs. Different bio-inspired solutions could be adopted for the deployment of WSNs with different goals, including robustness, maximum coverage and energy efficiency.

Algorithm	Coverage model	Coverage degree	D/C	R_c *vs.* R_s	Primary objective	Termination condition	Complexity	Communication overhead	Convergence speed	Sensor movement
ϵMSDVRG [WAN 06e]	B	S	C	$R_c = R_s$	Ensure the total coverage of the RoI	Stability state	High	Medium	Medium	Medium
CLP [WAN 06c]	B	S	D	$R_c \geq 2R_s$	Ensure a uniform coverage	Stability state	Medium	High	Slow	High
Isogrid [LAM 06]	B	S	D	$R_c > \sqrt{3}R_s$	Ensure a uniform coverage	Maximum number of iterations	Medium	Medium	Medium	High
S&S [BAR 08]	B	S	D	$R_c \geq \sqrt{3}R_s$	Ensure a uniform coverage	Stability state	Medium	High	Slow	High
P&P [BAR 10]	B	S	D	$R_c \geq \sqrt{3}R_s$	Ensure a uniform coverage	Stability state	Medium	High	Medium	Medium
Wang *et al*. [WAN 08b]	B & P	S	D	NDR	Ensure a uniform coverage	Stability state	High	Medium	Medium	Medium

Table 5.5. *Comparison between the coverage pattern based algorithms*

Algorithm	Coverage model	Coverage degree	D/C	R_c vs. R_s	Primary objective	Termination condition	Complexity	Communication overhead	Convergence speed	Sensor movement
SMART [YAN 07]	B	S	C	NDR	Balance the number of nodes in each cluster	Balanced load	Medium	High	Fast	High
SMART(g) , H-SMART [WU 07b], and SMART(m, c) [WU 07b]	B	S	C	NDR	Balance the number of nodes in each cluster	Balanced load	Medium	High	Fast	Medium
Cardei *et al.* [CAR 08a]	B	S	D	NDR	Increase the nodes density around the sink	Balanced load	Medium	Medium	Medium	Medium
Yang *et al.* [YAN 06b]	B	S	C	NDR	Balance the number of nodes in each cluster	Balanced load	High	Low	Fast	Low

Table 5.6. *Comparison between the grid quorum based algorithms*

Because the structure of organisms and networks is very alike, Nazi *et al.* [NAZ 14] investigated the use of topologies inspired by Gene Regulatory Networks (GRNs) for deploying WSNs. The authors considered multiple living organisms (i.e. yeast and E. coli) as templates for the proposed bio-inspired topology design. The goal is to deploy a WSN that is isomorphic to a subnetwork of the GRN. The authors exploited virtual forces to shape a WSN whose topology corresponded to a certain GRN sub-network. Results obtained, based on both simulations and experiments in a real WSN testbed, showed that bio-inspired WSNs are resilient to node and link failures, have low latency and are energy-efficient.

Nagchoudhury *et al.* [NAG 15] considered the regular hexagon to fully cover an area of interest. Sensors move to their optimal locations on the basis of the Bacteria Foraging Algorithm (BFA) that copies the searching behavior of *Escherichia coli* bacteria (E. Coli), which is in the small intestine of the human body. The deployed sensors are taken as the bacteria that are in search of the food which is depicted by the best possible communication link.

Table 5.7 provides a comparative summary of the characteristics of the bio-inspired algorithms.

5.2.8. *Miscellaneous approaches*

Our discussion so far has concerned mainly the deployment approach classes where several self-deployment algorithms share the same principle. However, in the literature, we found a few strategies that do not share their principle with the seven classes discussed earlier. In this section, we will discuss such strategies.

5.2.8.1. *Sensor deployment using self-organizing maps*

Koutsougeras *et al.* [KOU 08] tried to take into account the realistic consideration of the probability density for events to be sensed, termed as event-driven coverage. The objective is to distribute sensors so that the distribution density of the sensors matches that of the probability density of events to be sensed. In this context, they explored the concept of Self-Organizing Maps (SOMs) to address the coverage problem. The SOM is a method for unsupervised learning, based on a grid of artificial neurons whose weight vectors are adapted to match the distribution of vectors in a sample set topologically that is used as an exemplar or training set.

Algorithm	Coverage model	Coverage degree	D/C	R_c *vs.* R_s	Primary objective	Termination condition	Complexity	Communication overhead	Convergence speed	Sensor movement
Nazi *et al*. [NAZ 14]	B/P	S	C	NDR	Enhance network robustness	Stability state	High	Low	Medium	Medium
Nagchoudhury *et al*. [NAG 15]	B	S	D	NDR	Enhance coverage and connectivity	Maximum number of iterations	High	High	Slow	High

Table 5.7. *Comparison between the bio-inspired algorithms*

In the initialization stage, a set of sample points is generated to represent the distribution of events. The algorithm starts with an arbitrary initial placement of sensors, followed by an iterative adaptive process of rearrangement through the small movement of sensors. Each iteration focuses on one particular sample and the sensor that is closest to this sample is identified (based on the Euclidean distance). This sensor is called the "winner" for the current sample and this winner is allowed to move closer to the sample. This process is repeated until no more significant movement is attainable.

5.2.8.2. *Sensor deployment using market competition*

Inspired by the market competition of human society, Peng *et al.* [PEN 08] proposed a distributed algorithm to handle the wireless sensor network connectivity and coverage issues. In this algorithm, the sensors in the network are seen as enterprises in economic activities, the areas of interest are taken as resources, and network configurations are seen as market competitions. The entire network is divided into many independent sub-networks, which cannot communicate with each other. The network contains both static nodes that represent the large enterprise and have the advantage of occupying part of the resources, and mobile nodes that represent the small businesses where the non-allocated resources will be allocated based on the competition between them to construct the "quasi-static sensors coverage", the non-used dynamic sensors will participate in the expansion process.

5.2.8.3. *Sensor deployment using cellular learning automata*

Esnaashari *et al.* [ESN 11] proposed a self-deployment strategy based on cellular learning automata. Learning automata (LA) are adaptive decision-making devices that operate on unknown random environments. The automaton evolves to some final desired behavior by adapting its decisions based on the reinforcement signal delivered by the environment. Cellular learning automaton (CLA) is the combination of cellular automaton (CA) and learning automaton. The basic idea of CLA is to utilize LA to adjust the state transition of CA.

In the proposed algorithm [ESN 11], neighboring sensors apply forces to each other, which makes every sensor move according to the resultant force vector applied to it. Each sensor is equipped with a learning automaton that decides for the sensor, at any given time, whether to apply force on its neighbors or not. This way, each sensor gradually learns its best position within the RoI in order to ensure high coverage.

5.3. Sensor relocation algorithms

Many studies have exploited the mobility of sensors to improve the performance of WSNs. Examples of metrics considered by those studies include quality of coverage, connectivity, network lifetime and network resiliency. Movement-assisted sensor deployment algorithms discussed above strive to construct a WSN by distributing sensors in the RoI. This type of repositioning is pursued at the deployment phase and affects all the deployed sensors so that a better network topology is achieved and the network performance is increased. On the other hand, in sensor relocation, instead of relocating the sensors at the deployment phase, a limited number of sensors are relocated on demand to enhance certain performance metrics. This can be decided while the network is operational, based on the dynamic variations in user interests, network resources, and the surrounding environment.

Sensor relocation algorithms employ different approaches that are similar to those used by movement-assisted sensor deployment algorithms, such as virtual forces, grid quorum and fuzzy logic. As compared with sensor self-deployment, sensor relocation has many special difficulties. First, before relocating sensors, the sensor relocation protocol should localize the event that requires sensor relocation (dynamic variations generated by the network or the environment). Examples of such events include: sensor failure, energy resource depletion, coverage holes and broken links. In some cases, such as the healing of coverage holes, the sensor relocation protocol not only needs to detect and localize existing holes, but also estimate their sizes and shapes for efficient healing. Second, the relocation process should select only a subset of sensors that will be relocated as the response to those variations. The subset should be selected based on a set of efficiency and effectiveness metrics. Third, as sensor relocation is triggered as a response to an occurring event, in many cases a strict response time requirement is imposed. For instance, in mission-critical WSN applications such as surveillance, if some sensors start malfunctioning or run out of power, their replacement should be quick. Fourth, it is always crucial to preserve sensors' energy while dealing with the WSNs. If a sensor travels long distances to replace another failed sensor, it may run out of energy, thereby creating a new problem instead of resolving the earlier one. Thus, a good relocation scheme should be energy-aware. Finally, sensor relocation should not affect the application currently using the WSN, which means that the sensor relocation process should be both quick and lightweight to minimize its effect on the current topology and other applications in the network.

In order to achieve sensor relocation, at least three steps need to be performed. First, the event that requires sensor relocation should be localized. Second, a subset of sensors that will be relocated should be identified. Finally, the identified subset of sensors should be relocated. The energy consumption should be minimized during the process of finding the redundant sensors as well as when relocating them. Different movement pattern are possible such as direct and cascaded movement. In the latter, the idea is that moving intermediate sensors to the target location can reduce the delay and balance the energy consumption as compared to moving one sensor for a long distance.

Sensor relocation can be performed for different purposes: to heal coverage holes, restore connectivity, enhance target tracking, improve data delivery latency, perform network resource optimization, counteract security attacks, or for fault tolerance purposes. In the following sections, we further elaborate these techniques with examples from relevant research papers.

5.3.1. *Sensor relocation for hole healing*

Approaches within this category deal with hole healing. The event of interest that requires sensor relocation is the emergence of holes in the RoI, which is often unavoidable due to the inner nature of WSNs, random deployment, environmental factors and external attacks. The basic idea of these approaches is to detect uncovered areas (holes) and relocate some sensors to fill them. Of course, this should be done while avoiding the creation of new holes. Other metrics such as network connectivity could be considered. To ensure effective hole healing, four key elements were identified by Senouci *et al.* [SEN 13b], namely: (i) determining the boundary of the RoI, (ii) detecting coverage holes and estimating their characteristics, (iii) determining the best target locations to relocate sensors to repair holes, and (iv) dispatching sensors to the target locations while minimizing the moving and messaging cost.

There has been much related research on the hole and border detection problem [GHR 05, KUN 06, SEN 13b]. Proposed approaches exploit different types of information such as connectivity information [GHR 05], the topology of the communication graph [KUN 06] or geographic information [SEN 13b].

For instance, in [KUN 06] a sensor decides whether it is on the boundary of a hole by comparing its degree with the average degree of its 2-hop neighbors. Once the holes are identified and characterized, sensors should be relocated to the best target locations. For instance, Senouci *et al.* [SEN 13b] exploit virtual forces to heal the discovered holes. They define the hole healing area in which the forces will be effective. This allows a local healing where only the sensors located at an appropriate distance from the hole will be involved in the healing process. In [IZA 15], the authors use a Type-2 fuzzy logic system (FLS2) to choose the sensors to be moved.

5.3.2. *Sensor relocation for connectivity restoration*

In many real-world WSN applications, collaboration among the sensors is primordial. In this case, maintaining network connectivity is critical to the effectiveness of WSNs. While connectivity can be ensured at the deployment phase, sudden sensor failures can break communication links. Therefore, restoring connectivity is the primary objective of many sensor relocation schemes [YOU 10, TAM 10, LEE 10, JOS 13, JOS 15]. The operating principal of these schemes is detecting a sensor failure and replacing the failed sensor with one or many of its neighbors to restore connectivity in the network. For instance, Younis *et al.* [YOU 10] proposed that sensors periodically send heartbeat messages to their neighbors, and each sensor maintains a list of its 1-hop neighbors. A sensor assumes the failure of a neighbor sensor s_f upon the missing of the s_f heartbeat messages. The basic idea of the recovery scheme is moving the 1-hop neighbors of s_f inward toward the position of s_f until they become $R_c/2$ away from the position of s_f. Such a move gets all neighbors of s_f connected [YOU 10].

In contrast to the above-mentioned relocation process that completely ignores the coverage issue, other connectivity restorations are coverage-aware such as C^3R and ECR [TAM 10], they consider both coverage and connectivity. In addition, the aforementioned schemes only deal with a single sensor failure or multiple non-collocated failures. The failure of multiple collocated sensors has been considered in [LEE 10, JOS 13, JOS 15]. These approaches use a common meeting point or exploit the pre-failure network topology to restore connectivity in the network.

5.3.3. *Sensor relocation for network resource optimization*

The objective of some sensor relocation schemes is network resource optimization such as prolonging the network lifetime through relocating existing mobile sensors. For instance, authors in [WAN 10b] consider sensor relocation in a cluster-based mobile WSN. They propose a centralized relocation scheme that relocates redundant sensors to the low battery energy clusters to increase the overall WSN lifetime while ensuring coverage. Redundant sensors are selected from clusters having high battery energy. In [COS 08], the author proposed a sensor relocation scheme *MaxNetLife* that aims at extending the WSN lifetime. The main idea is to relocate redundant sensors to the locations where the power consumption rate is high.

5.3.4. *Sensor relocation for fault tolerance purposes*

In order to mitigate failures in WSNs, tolerance mechanisms are usually employed [CHO 15]. They can be classified into two categories: proactive (preventive) or reactive (curative). The former deploys redundant sensors at the network setup phase in order to provide the network with fault-tolerance properties such as *k*-coverage and *k*-connectivity. The latter provides the network with self-healing capabilities by exploiting sensors' mobility.

Self-healing schemes are usually two-phase solutions: redundant sensors are first identified and then relocated to the target location [YOU 14]. In [WAN 05a], the nearest redundant sensors are first identified using a grid-quorum approach, and then relocated to the target location using a cascaded movement. A similar approach is proposed in [CHE 15]. The proposed algorithm (*ETSR*) uses a grid-quorum approach to detect redundant sensors, and moves them to faulty sensors and coverage holes by cascaded movement. It also considers the estimated value of network lifetime in the planning of cascading schedule. In [LI 06], the authors proposed *ZONER*, a sensor relocation protocol based on a restricted flooding technique. *ZONER* discovers previously deployed redundant sensors and relocates them in a shifting way to replace failed non-redundant ones without changing the network topology. It should be noted that other relocation approaches such as those that target the restoration of coverage and/or connectivity could also be included in this category.

5.4. Practical issues that need further research

The effectiveness of adding mobility to wireless sensor networks has been shown to be certain. Indeed, mobile WSNs are resilient to failures, and reactive to dynamic variations in user interests, network resources and the surrounding environment. However, there is still vast scope for future work in this field. This section highlights the major challenges and open problems for future research work.

5.4.1. *Security vulnerabilities*

Security vulnerabilities that are specific to self-deployment and relocation algorithms is a serious issue that needs to be considered in future research. For instance, Bartolini *et al.* [BAR 14b] introduced an attack, called *opportunistic movement*, and showed that in a typical scenario this attack could reduce coverage by more than 50%, by compromising only 7% of the sensor nodes. The authors also show how to secure virtual forces and Voronoi-based deployment algorithms [BAR 14b, BAR 14a]. Future research in this area, that still remains vastly unexplored, should consider potential security vulnerabilities.

5.4.2. *WSN self-deployment and self-healing in three-dimensional RoI*

The scope of the majority of published papers on mobile WSN deployment is limited to two-dimensional RoI. However, with the growing interest in three-dimensional (3D) application scenarios such as underwater surveillance [SEN 13a], tackling the self-deployment and sensor relocation issues in 3D has become a necessity. Among the problems raised, the following issues are of particular interest: how can we model sensors, sensing capability and quality in 3D, and how can we incorporate topological information in the deployment process? Answers to these questions provided important insights into the design and performance of 3D WSNs.

In the literature, two variants of the 3D WSN coverage problem were considered: 3D surface [JIN 12] (e.g. mountain surface deployment) and 3D volume [SEN 13a] (e.g. underwater sensor deployment). To tackle the coverage issue in 3D, several solutions were proposed such as reducing the geometric problem from a 3D space to a 2D space [HUA 04], or abstracting the 3D RoI coverage as the ball-coverage problem, that is, covering a 3D space with several numbers of balls with the same radius [AMM 10, ZHA 10, XIA 14]. The former approach is not suitable to WSN self-deployment, whereas the latter solution remains very limited with little practical interest.

Though there are many works on the 3D static WSN coverage problem, few studies focus on the 3D WSN self-deployment problem [LI 13, MIA 15, CHE 14b, CHE 14a]. In [LI 13, MIA 15], the authors adapt the traditional VFA to 3D space, whereas Chen and Qian [CHE 14b, CHE 14a] exploited the fluid dynamics approach, where the WSN is viewed as a fluid body while nodes as charged particles. We envision WSN self-deployment and sensor relocation in 3D to be a promising research direction. Moreover, connectivity is still an open interesting problem in 3D deployment domains.

5.4.3. *Path planning*

When looking at the sensor nodes' movement from the perspective of robotics, a key problem is: how to plan the motion of mobile sensors? In other words, we have to deal with the path planning problem consisting in finding paths on which these mobile sensors can move to desired destinations while avoiding obstacles.

In the context of WSNs, path planning was mainly investigated for planning the motion of data mules that collect the data from stationary sensor nodes [SUG 11]. A possible criticism to almost all above-mentioned sensor deployment and sensor relocation algorithms is the strong assumption they make on the capabilities of the mobile sensor nodes, in particular the ability of the sensor nodes to move to any location where they are asked to move. When obstacles must be overcome, a great negative impact on energy consumption should be expected on all self-deployment protocols and sensor relocation schemes. Thus, future research in this area should consider path planning at the design stage of deployment approaches.

5.4.4. *Application-specific WSN self-deployment and self-healing*

Although the deployment of WSNs is usually application-driven, most of the current self-deployment strategies, and sensor relocation schemes do not include either the user interests or the application needs.

For instance, the deployment of wireless multimedia sensor networks (WMSNs) requires not only the overall coverage of the RoI, but also the sensors need to be placed not too close to each other to avoid or minimize communication interferences between them. This condition is very important to ensure the correct functioning of the multi-path routing, which is extensively used in this kind of network. This may require a self-deployment approach that is radically different from the aforementioned approaches.

We expect application-specific WSN self-deployment and self-healing to gain increased attention with the growing list of WSN applications and the availability of sensor-specific models of sensors' capabilities.

5.5. Conclusion

Adding mobility to wireless sensor networks enables numerous functionalities such as self-deployment, self-healing and adaptive resource management. Future works should consider self-deployment and self-healing under security constraints. They should also consider path planning at the design stage of deployment approaches. As we indicated, deployment of mobile WSNs in 3D will require increased attention from the research community in order to tackle practical deployment scenarios. Finally, further research on application-specific mobile WSN deployment can provide valuable information to optimize network resources in a sensing field according to the application needs, the surrounding environment and the user interests.

Conclusion

The last decade has witnessed rapid advancements in the development of wireless sensor networks (WSNs). Although WSNs promise unprecedented opportunities for monitoring and controlling our physical surroundings, there are several challenges that need to be addressed in order to deliver the potential advantages of this technology. The deployment of WSNs, which is the focus of this book, is one of the most important issues to be dealt with. Indeed, WSN deployment is of utmost importance in the process of developing WSN solutions for real-life applications, as it is a deciding factor of required resources and their configuration.

In this book, we have attempted to provide a comprehensive guide of fundamental concepts, new ideas, results, and open research issues in the area of WSN deployment. In Chapter 1, we introduced the background of WSNs and their deployment. We also elaborated on the analytical models concerning sensor coverage and communication. Then, in Chapter 2, we focused on the most *naive* approach to deploy WSNs, namely: random deployment. The latter is often the best choice when considering inaccessible or harsh environments and/or large-scale WSNs. In Chapter 3, we studied the WSNs deterministic deployment, which aims at generating a flat network topology that satisfies user's requirements. We analyzed the uncertainty-aware WSNs deployment where sensors may not always provide reliable information, and showed how the evidence theory could be exploited to design better deployment strategies. Chapter 4 was devoted to investigating the deployment of WSNs for critical applications requiring high detection rates coupled with low false alarms. To deal with dynamic variations in user interests, network resources and surrounding environments, we considered the deployment of mobile wireless sensor networks in Chapter 5. In this chapter, we showed that exploiting the mobility of mobile sensors wisely can enable numerous functionalities such as self-deployment, self-healing, and adaptive resource management.

In summary, this book discussed both theoretical and practical aspects, and provided guidelines for effective deployment of WSNs. To conclude this book, we summarized the major challenges and open problems for future research work on WSNs deployment.

For random deployment, more realistic sensor coverage and communication models should be considered. Indeed, there is an unquestionable need for realistic coverage and radio models that effectively characterize the effects of random deployment, while being concise enough to promote strong theoretical results. The effects of random deployment, such as heterogeneous sensor postures, should also be considered while designing sensor packages. Future research in this area should also consider investigating random deployment of WSNs in three-dimensional spaces, along with extending the asymptotic results for WSNs with practical sizes. For deterministic and fusion-based deterministic deployment, approaches addressing practical constraints are a major open problem in this field. The problem of sensor placement should be considered in more realistic settings, regarding all deployment-related aspects, such as sensors heterogeneity and reliability, placement errors, and environmental factors. Furthermore, future work should study the WSNs deployment problem not only from a computer science efficiency perspective, but should also focus on specific-application requirements by emphasizing cross-disciplinary collaboration. Finally, a set of research issues concerning the deployment of mobile wireless sensor networks needs to be addressed. Those include security requirements and path planning. Energy-efficient self-deployment and self-healing solutions are also required.

In summary, future research work on WSNs deployment should consider more realistic settings, while focusing on application demands, under joint constraints of WSNs, terrain, and other environmental factors.

Bibliography

[ABA 09a] Ababnah A., Natarajan B., "Optimal sensor deployment for value-fusion based detection", *IEEE Global Telecommunications Conference (GLOBECOM'09)*, Honolulu, HI, pp. 1–6, 2009.

[ABA 09b] Ababnah A., Natarajan B., "Sensor deployment as an optimal control problem", *Proceedings of the 18th International Conference on Computer Communications and Networks (ICCCN'09)*, Washington, DC, USA, pp. 1–5, 2009.

[ABA 10a] Ababnah A., Natarajan B., "LQR formulation of sensor deployment for decision fusion based detection", *IEEE Global Telecommunications Conference (GLOBECOM'10)*, Miami, FL, pp. 1–5, 2010.

[ABA 10b] Ababnah A., Sensor deployment in detection networks – a control theoretic approach, PhD Thesis, Kansas State University, 2010.

[ABA 11] Ababnah A., Natarajan B., "Optimal control-based strategy for sensor deployment", *IEEE Transactions on Systems, Man and Cybernetics, Part A: Systems and Humans*, vol. 41, no. 1, pp. 97–104, 2011.

[ABO 15] Abo-Zahhad M., Ahmed S.M., Sabor N. *et al.*, "Rearrangement of mobile wireless sensor nodes for coverage maximization based on immune node deployment algorithm", *Computers and Electrical Engineering*, vol. 43, pp. 76–89, 2015.

[AFS 14] Afsar M.M., Tayarani-N M.-H., "Clustering in sensor networks: a literature survey", *Journal of Network and Computer Applications*, vol. 46, pp. 198–226, 2014.

[AIT 07] Aitsaadi N., Achir N., Boussetta K. *et al.*, "Differentiated underwater sensor network deployment", *EEE/OES, OCEANS'07*, pp. 1–6, 2007.

[AIT 09] Aitsaadi N., Achir N., Boussetta K. *et al.*, "A tabu search WSN deployment method for monitoring geographically irregular distributed events", *Sensors*, vol. 9, pp. 1625–1643, 2009.

[AIT 11] Aitsaadi N., Achir N., Boussetta K. *et al.*, "Artificial potential field approach in WSN deployment: cost, QoM, connectivity, and lifetime constraints", *Computer Networks*, vol. 55, no. 1, pp. 84–105, January 2011.

[AKB 13] AKBARZADEH V., GAGNÉ C., PARIZEAU M. *et al.*, "Probabilistic sensing model for sensor placement optimization based on line-of-sight coverage", *IEEE Transactions on Instrumentation and Measurement*, vol. 62, no. 2, pp. 293–303, February 2013.

[AKK 07] AKKAYA K., YOUNIS M., "C2AP: coverage-aware and connectivity-constrained actor positioning in wireless sensor and actor networks", *IEEE International Performance, Computing, and Communications Conference (IPCCC'07)*, New Orleans, LA, pp. 281–288, 2007.

[AKY 02] AKYILDIZ I.F., SU W., SANKARASUBRAMANIAM Y. *et al.*, "Wireless sensor networks: a survey", *Computer Networks*, vol. 38, no. 4, pp. 393–422, March 2002.

[AKY 10] AKYILDIZ I.F., VURAN M.C., *Wireless Sensor Networks*, Wiley, 2010.

[ALA 15] ALAM S.N., HAAS Z.J., "Coverage and connectivity in three-dimensional networks with random node deployment", *Ad Hoc Networks*, vol. 34, pp. 157–169, 2015.

[ALS 15] ALSKAIF T., ZAPATA M.G., BELLALTA B., "Game theory for energy efficiency in wireless sensor networks: latest trends", *Journal of Network and Computer Applications*, vol. 54, pp. 33–61, 2015.

[AMA 12] AMALDI E., CAPONE A., CESANA M. *et al.*, "Design of wireless sensor networks for mobile target detection", *IEEE/ACM Transactions on Networking*, vol. 20, no. 3, pp. 784–797, 2012.

[AMM 08] AMMARI H.M., DASS S.K., "Integrated coverage and connectivity in wireless sensor networks: a two-dimensional percolation problem", *IEEE Transactions on Computers*, vol. 57, no. 10, pp. 1423–1434, 2008.

[AMM 10] AMMARI H., DAS S., "A study of k-coverage and measures of connectivity in 3D wireless sensor networks", *IEEE Transactions on Computers*, vol. 59, no. 2, pp. 243–257, 2010.

[AMM 14] AMMARI H.M. (ed.), *The Art of Wireless Sensor Networks*, vol. 2, Springer, 2014.

[APP 91] APPRIOU A., "Formulation et traitement de l'incertain en analyse multisenseurs", *GRETSI*, pp. 951–954, 1991.

[ARD 15] ARDUINO, "What is Arduino?", available at: http://www.arduino.cc/, 2015.

[ARO 04] ARORA A., DUTTA P., BAPAT S. *et al.*, "A line in the sand: A wireless sensor network for target detection, classification, and tracking", *Computer Networks*, vol. 46, no. 5, pp. 605–634, 2004.

[ATT 12] ATTEA B., OKAY F., OZDEMIR S. *et al.*, "Multi-objective evolutionary algorithm based on decomposition for efficient coverage control in mobile sensor networks", *6th International Conference on Application of Information and Communication Technologies (AICT)*, Tbilisi, pp. 1–6, October 2012.

[AZI 09] AZIZ N.A.A., AZIZ K.A., ISMAIL W.Z.W., "Coverage strategies for wireless sensor networks", *World Academy of Science, Engineering and Technology*, vol. 3, no. 2, pp. 134–140, 2009.

[BAH 09] BAHAREH J., FARAHANI H.G., FATHY M., "A fuzzy based priority approach in mobile sensor network coverage", *International Journal of Recent Trends in Engineering*, vol. 2, no. 1, pp. 138–143, November 2009.

[BAL 09] BALISTER P., KUMAR S., "Random vs. deterministic deployment of sensors in the presence of failures and placement errors", *IEEE INFOCOM 2009 Proceedings*, Rio de Janeiro, pp. 2896–2900, 2009.

[BAR 08] BARTOLINI N., CALAMONERI T., FUSCO E.G. *et al.*, "Snap and spread: a self-deployment algorithm for mobile sensor networks", *Proceedings of the 4th IEEE international conference on Distributed Computing in Sensor Systems*, Berlin, Heidelberg, Springer-Verlag, pp. 451–456, 2008.

[BAR 10] BARTOLINI N., CALAMONERI T., FUSCO E.G. *et al.*, "Push & Pull: autonomous deployment of mobile sensors for a complete coverage", *Wireless Networks*, Kluwer Academic Publishers, vol. 16, no. 3, pp. 607–625, 2010.

[BAR 14a] BARTOLINI N., BONGIOVANNI G., PORTA T.L. *et al.*, "Voronoi-based deployment of mobile sensors in the face of adversaries", *IEEE International Conference on Communications (ICC)*, Sydney, NSW, pp. 532–537, 2014.

[BAR 14b] BARTOLINI N., BONGIOVANNI G., PORTA T.L. *et al.*, "On the vulnerabilities of the virtual force approach to mobile sensor deployment", *IEEE Transactions on Mobile Computing*, vol. 13, no. 11, pp. 2592–2605, November 2014.

[BHU 12] BHUIYAN M.Z.A., WANG G., CAO J., "Sensor placement with multiple objectives for structural health monitoring in WSNs", *14th IEEE International Conference on High Performance Computing and Communications (HPCC)*, Liverpool, pp. 699–706, 2012.

[BUE 06] BUETTNER M., YEE G.V., ANDERSON E. *et al.*, "X-MAC: a short preamble MAC protocol for duty-cycled wireless sensor networks", *Proceedings of the 4th International Conference on Embedded Networked Sensor Systems (SenSys'06)*, Boulder, Colorado, USA, pp. 307–320, 2006.

[CAR 08a] CARDEI M., YANG Y., WU J., "Non-uniform sensor deployment in mobile wireless sensor networks", *International Symposium on a World of Wireless, Mobile and Multimedia Networks (WoWMoM)*, Newport Beach, CA, pp. 1–8, 2008.

[CAR 08b] CARTER B., RAGADE R., "An extensible model for the deployment of non-isotropic sensors", *IEEE Sensors Applications Symposium*, Atlanta, GA, pp. 22–25, 2008.

[CHA 86] CHAIR Z., VARSHNEY P., "Optimal data fusion in multiple sensor detection systems", *IEEE Trans. Aerosp. Electron. Syst*, vol. AES-22, no. 1, pp. 98–101, 1986.

[CHA 02] CHAKRABARTY K., IYENGAR S.S., QI H. *et al.*, "Grid coverage for surveillance and target location in distributed sensor networks", *IEEE Trans. Comput.*, IEEE Computer Society, vol. 51, pp. 1448–1453, December 2002.

[CHA 11] CHANG X., TAN R., XING G. *et al.*, "Sensor placement algorithms for fusion-based surveillance networks", *IEEE Trans. Parallel Distrib. Syst.*, vol. 22, pp. 1407–1414, August 2011.

[CHE 02] CHEN A., MUNTZ R., YUEN S. *et al.*, "A support infrastructure for the smart kindergarten", *IEEE Pervasive Computing*, vol. 1, no. 2, pp. 49–57, 2002.

[CHE 07a] CHELLAPPAN S., BAI X., MA B. *et al.*, "Mobility limited flip-based sensor networks deployment", *IEEE Trans. Parallel Distrib. Syst.*, vol. 18, pp. 199–211, February 2007.

[CHE 07b] CHEN J., LI S., SUN Y., "Novel deployment schemes for mobile sensor networks", *Sensors*, vol. 7, pp. 2907–2919, 2007.

[CHE 09] CHEN J., SHEN E., SUN Y., "The deployment algorithms in wireless sensor netWorks: a survey", *Information Technology Journal*, vol. 8, no. 3, pp. 293–301, 2009.

[CHE 14a] CHEN J., QIAN H., "Node deployment algorithm based on viscous fluid model for wireless sensor networks", *The Scientific World Journal*, vol. 2014, p. 8, 2014.

[CHE 14b] CHEN J., QIAN H., "A novel deployment algorithm based on fluid dynamics approach", *International Journal of Smart Home*, vol. 8, no. 4, pp. 83–96, 2014.

[CHE 14c] CHEN J., TAN R., WANG Y. *et al.*, "A sensor system for high-fidelity temperature distribution forecasting in data centers", *ACM Trans. Sensor Netw.*, vol. 11, no. 2, p. 25, December 2014.

[CHE 15] CHENG C.-F., HUANG C.-W., "An energy-balanced and timely self-relocation algorithm for grid-based mobile WSNs", *IEEE Sensors Journal*, vol. 15, no. 8, pp. 4184–4193, August 2015.

[CHO 15] CHOUIKHI S., KORBI I.E., GHAMRI-DOUDANE Y. *et al.*, "A survey on fault tolerance in small and large scale wireless sensor networks", *Computer Communications*, vol. 69, pp. 22–37, September 2015.

[CLO 02] CLOUQUEUR T., PHIPATANASUPHORN V., RAMANATHAN P. *et al.*, "Sensor deployment strategy for target detection", *Proceedings of the 1st ACM international workshop on Wireless sensor networks and applications (WSNA'02)*, New York, USA, pp. 42–48, 2002.

[CLO 03] CLOUQUEUR T., PHIPATANASUPHORN V., RAMANATHAN P. *et al.*, "Sensor deployment strategy for detection of targets traversing a region", *Mob. Netw. Appl.*, Kluwer Academic Publishers, vol. 8, pp. 453–461, August 2003.

[CLO 04] CLOUQUEUR T., SALUJA K.K., RAMANATHAN P., "Fault tolerance in collaborative sensor networks for target detection", *IEEE Transactions on Computers*, vol. 53, pp. 320–333, March 2004.

[COE 02] COELLO C., "Theoretical and numerical constraint-handling techniques used with evolutionary algorithms: a survey of the state of the art", *Comp. Methods in Applied Mechanics and Engineering*, vol. 191, pp. 1245–1287, 2002.

[COS 08] COSKUN V., "Relocating sensor nodes to maximize cumulative connected coverage in wireless sensor networks", *Sensors*, vol. 8, pp. 2792–2817, 2008.

[DEB 02] DEB K., PRATAP A., AGARWAL S. *et al.*, "A fast and elitist multiobjective genetic algorithm: NSGA-II", *IEEE Transactions on Evolutionary Computation*, vol. 6, no. 2, pp. 182–197, 2002.

[DHI 02] DHILLON S.S., CHAKRABARTY K., IYENGAR S.S., "Sensor placement for grid coverage under imprecise detections", *Proc. Fifth Int Information Fusion Conf*, vol. 2, pp. 1581–1587, 2002.

[DHI 03] DHILLON S.S., CHAKRABARTY K., "Sensor placement for effective coverage and surveillance in distributed sensor networks", *Proc. IEEE Wireless Communications and Networking (WCNC)*, vol. 3, pp. 1609–1614, 2003.

[DRE 10] DRESSLER F., AKAN O., "Bio-inspired networking: from theory to practice", *IEEE Commun. Mag.*, vol. 48, no. 11, pp. 176–183, 2010.

[DU 07] DU X., ZHANG M., NYGARD K.E. *et al.*, "Self-healing sensor networks with distributed decision making", *International Journal of Sensor Networks*, Inderscience Publishers, vol. 2, no. 5/6, pp. 289–298, July 2007.

[DU 15] DU W., XING Z., LI M. *et al.*, "Sensor placement and measurement of wind for water quality studies in urban reservoirs", *ACM Transactions on Sensor Networks*, vol. 11, no. 3, pp. 41–68, February 2015.

[DUA 04] DUARTE M.F., HU Y.H., "Vehicle classification in distributed sensor networks", *J. Parallel Distrib. Comput.*, Academic Press, Inc., vol. 64, no. 7, pp. 826–838, July 2004.

[DYO 10] DYO V., S.ELLWOOD, MACDONALD D. *et al.*, "Evolution and sustainability of a wildlife monitoring sensor network", *Proceedings of the 8th ACM Conference on Embedded Networked Sensor Systems (SenSys'10)*, New York, ACM, pp. 127–140, 2010.

[ELO 04] ELOUEDI Z., MELLOULI K., SMETS P., "Assessing sensor reliability for multisensor data fusion within the transferable belief model", *IEEE Transactions on Systems, Man, and Cybernetics*, vol. 34, no. 1, pp. 782–787, 2004.

[ESL 13] ESLAMI A., NEKOUI M., PISHRO-NIK H. *et al.*, "Results on finite wireless sensor networks: connectivity and coverage", *ACM Transactions on Sensor Networks*, vol. 9, no. 4, p. 22, July 2013.

[ESN 11] ESNAASHARI M., MEYBODI M., "A cellular learning automata-based deployment strategy for mobile wireless sensor networks", *J. Parallel Distrib. Comput.*, vol. 71, no. 7, pp. 988–1001, 2011.

[FAN 14] FAN T., TENG G., HUO L., "A pre-determined nodes deployment strategy of two-tiered wireless sensor networks based on minimizing cost", *Int. J. Wireless Inf. Networks*, vol. 21, no. 2, pp. 114–124, 2014.

[FEI 09] FEI X., BOUKERCHE A., ARAUJO R.B., "Irregular sensing range detection model for coverage based protocols in wireless sensor networks", *Proceedings of the 28th IEEE Conference on Global Telecommunications (GLOBECOM'09)*, Piscataway, NJ, USA, pp. 5699–5704, 2009.

[FIR 15] FIRESENSE, "FIRESENSE Project", available at: http://www.firesense.eu/, 2015.

[FOR 97] FORTUNE S., "Voronoi diagrams and Delaunay triangulations", in TOTH C.D., O'ROURKE J., GOODMAN J.E. (eds), *Handbook of Discrete and Computational Geometry*, CRC Press, Boca Raton, FL, 1997.

[GEO 13] GEORGE J., KAPLAN L.M., "Shooter localization using a wireless sensor network of soldier-worn gunfire detection systems", *Journal of Advances in Information Fusion*, vol. 8, no. 1, pp. 15–32, June 2013.

[GHO 06] GHOSH A., DAS S.K., "Coverage and connectivity issues in wireless sensor networks", in SHOREY R. *et al.* (eds), *Mobile, Wireless and Sensor Networks*, John Wiley & Sons, 2006.

[GHR 05] GHRIST R., MUHAMMAD A., "Coverage and hole-detection in sensor networks via homology", *Proceedings of IPSN'05*, UCLA, Los Angeles, pp. 254–260, April 2005.

[GU 05] GU L., JIA D., VICAIRE P. *et al.*, "Lightweight detection and classification for wireless sensor networks in realistic environments", *Proceedings of the 3rd international conference on Embedded networked sensor systems (SenSys'05)*, San Diego, California, USA, pp. 205–217, 2005.

[GUP 98] GUPTA P., KUMAR P.R., "Critical power for asymptotic connectivity in wireless networks", in MCENEANEY W.M., YIN G., ZHANG Q. (eds), *Stochastic Analysis, Control, Optimization and Applications: A Volume in Honor of W.H. Fleming*, Springer, 1998.

[HAL 15] HALDER S., BIT S.D., "Design of an Archimedes' spiral based node deployment scheme targeting enhancement of network lifetime in wireless sensor networks", *Journal of Network and Computer Applications*, vol. 47, pp. 147–167, 2015.

[HE 06] HE T., KRISHNAMURTHY S., LUO L. *et al.*, "VigilNet: an integrated sensor network system for energy-efficient surveillance", *ACM Trans. Sen. Netw.*, vol. 2, pp. 1–38, February 2006.

[HEO 03] HEO N., VARSHNEY P., "A distributed self spreading algorithm for mobile wireless sensor networks", *IEEE Wireless Communications and Networking (WCNC)*, vol. 3, New Orleans, LA, USA, pp. 1597–1602, 2003.

[HOW 02] HOWARD A., MATARIC M.J., SUKHATME G.S., "Mobile sensor network deployment using potential fields: a distributed, scalable solution to the area coverage problem", *Proceedings of the 6th International Symposium on Distributed Autonomous Robotics Systems (DARS'02)*, Fukuoka, Japan, pp. 299–308, 2002.

[HUA 04] HUANG C., TSENG Y.-C., LO L.-C., "The coverage problem in three-dimensional wireless sensor networks", *IEEE Global Telecommunications Conference (GLOBECOM'04)*, vol. 5, pp. 3182–3186, November 2004.

[HUA 13] HUANG C., CHANG H.-Y., WU K.-L., "A jigsaw-based sensor placement algorithm for wireless sensor networks", *International Journal of Distributed Sensor Networks*, vol. 2013, p. 11, 2013.

[IQB 04] IQBAL M., GONDAL I., DOOLEY L., "Dynamic symmetrical topology models for pervasive sensor networks", *Proceedings of INMIC*, 2004.

[ISH 04a] ISHIZUKA M., AIDA M., "Performance study of node placement in sensor networks", *International Conference on Distributed Computing Systems Workshops*, vol. 5, Los Alamitos, CA, USA, pp. 598–603, 2004.

[ISH 04b] ISHIZUKA M., AIDA M., "The reliability performance of wireless sensor networks configured by power-law and other forms of stochastic node placement", *IEICE Trans. on Communications*, vol. E87-B, no. 9, pp. 2511–2520, September 2004.

[IZA 15] IZADI D., ABAWAJY J., GHANAVATI S., "An alternative node deployment scheme for WSNs", *IEEE Sensors Journal*, vol. 15, no. 2, pp. 667–675, February 2015.

[JIA 08] Jiang X., Chen Y., Yu T., "Localized distributed sensor deployment via coevolutionary computation", *Third International Conference on Communications and Networking in China (ChinaCom)*, Hangzhou, pp. 785–789, 2008.

[JIN 12] Jin M., Rong G., Wu H. *et al.*, "Optimal surface deployment problem in wireless sensor networks", *Proceedings of IEEE INFOCOM*, USA, pp. 2345–2353, March 2012.

[JOS 13] Joshi Y., Younis M., "Distributed approach for reconnecting disjoint segments", *Proceedings of IEEE Global Communications Conference (GLOBECOM'13)*, Atlanta, GA, pp. 255–260, December 2013.

[JOS 15] Joshi Y.K., Younis M., "Exploiting skeletonization to restore connectivity in a wireless sensor network", *Computer Communications*, vol. 99, pp. 1–11, 2015.

[JUU 15] Juul J.P., Green O., Jacobsen R.H., "Deployment of wireless sensor networks in crop storages", *Wireless Personal Communications*, vol. 81, no. 4, pp. 1437–1454, 2015.

[KE 11] Ke W.-C., Liu B.-H., Tsai M.-J., "The critical-square-grid coverage problem in wireless sensor networks is NP-Complete", *Computer Networks*, vol. 55, no. 9, pp. 2209–2220, 2011.

[KEA 14] Keally M., Zhou G., Xing G. *et al.*, "A learning-based approach to confident event detection in heterogeneous sensor networks", *ACM Transactions on Sensor Networks*, vol. 11, no. 1, p. 28, November 2014.

[KER 39] Kershner R., "The number of circles covering a set", *American Journal of Mathematics*, vol. 61, no. 3, pp. 665–671, 1939.

[KHA 15] Khan J.A., Qureshi H.K., Iqbal A., "Energy management in wireless sensor networks: a survey", *Computers and Electrical Engineering*, vol. 41, pp. 159–176, 2015.

[KIM 08] Kim Y., Schmid T., Charbiwala Z.M. *et al.*, "NAWMS: nonintrusive autonomous water monitoring system", *Proceedings of the 6th ACM Conference on Embedded Network Sensor Systems (SenSys'08)*, NC, USA, pp. 309–322, November 2008.

[KOU 08] Koutsougeras C., Liu Y., Zheng R., "Event-driven sensor deployment using self-organizing maps", *International Journal of Sensor Networks*, Inderscience Publishers, vol. 3, no. 3, pp. 142–151, May 2008.

[KUM 04] Kumar S., Lai T.H., Balogh J., "On k-coverage in a mostly sleeping sensor network", *Proceedings of the 10th Annual International Conference on Mobile Computing and Networking (MobiCom'04)*, Philadelphia, PA, USA, pp. 144–158, 2004.

[KUN 06] Kun B., T.K., G.N. *et al.*, "Topological hole detection in sensor networks with cooperative neighbors", *International Conference on Systems and Networks Communications (ICSNC'06)*, Los Alamitos, CA, USA, p. 31, 2006.

[LAI 05] Laibowitz M., Paradiso J., "Parasitic mobility for pervasive sensor networks", *Proceedings of the third International Conference on Pervasive Computing (PERVASIVE'05)*, Berlin, Heidelberg, pp. 255–278, 2005.

[LAI 12] Lai T.T., Hen W., Li K. *et al.*, "TriopusNet: automating wireless sensor network deployment and replacement in pipeline monitoring", *Proceedings of the 11th International Conference on Information Processing in Sensor Networks (IPSN'12)*, Beijing, China, pp. 61–71, April 16–20 2012.

[LAM 06] LAM M.L., LIU Y.H., "ISOGRID: an efficient algorithm for coverage enhancement in mobile sensor networks", *Proceedings of the 2006 IEEE/RSJ International Conference on Intelligent Robots and Systems*, Beijing, China, pp. 1458–1463, October 2006.

[LAU 57] LAURENT A., "Bombing problems – a statistical approach", *Operations Research*, vol. 5, no. 1, pp. 75–89, 1957.

[LE 15] LE D.V., OH H., YOON S., "VirFID: a virtual force (VF)-based interest-driven moving phenomenon monitoring scheme using multiple mobile sensor nodes", *Ad Hoc Networks*, vol. 27, pp. 112–132, 2015.

[LEE 10] LEE S., YOUNIS M., "Recovery from multiple simultaneous failures in wireless sensor networks using minimum Steiner tree", *J. Parallel Distrib. Comput.*, vol. 70, no. 5, pp. 525–536, April 2010.

[LI 05] LI S., XU C., PAN W. *et al.*, "Sensor deployment optimization for detecting maneuvering targets", *8th International Conference on Information Fusion*, vol. 2, Philadelphia, PA, USA, pp. 1629–1635, 2005.

[LI 06] LI X., SANTORO N., "ZONER: a ZONE-based sensor relocation protocol for mobile sensor networks", *Proceedings of the 31st IEEE Conference on Local Computer Networks*, Tampa, FL, pp. 923–930, November 2006.

[LI 10] LI B., WANG D., WANG F. *et al.*, "High quality sensor placement for SHM systems: refocusing on application demands", *IEEE INFOCOM Proceedings*, San Diego, CA, pp. 1–9, 2010.

[LI 11] LI M., CHENG W., LIU K. *et al.*, "Sweep coverage with mobile sensors", *IEEE Transactions on Mobile Computing*, vol. 10, no. 11, pp. 1534–1545, 2011.

[LI 13] LI X., CI L., YANG M. *et al.*, "Deploying three-dimensional mobile sensor networks based on virtual forces algorithm", WANG R., XIAO F. ((eds.)), *Advances in Wireless Sensor Networks*, vol. 334, pp. 204–216, Springer Berlin Heidelberg, 2013.

[LIA 15] LIAO Z., WANG J., ZHANG S. *et al.*, "Minimizing movement for target coverage and network connectivity in mobile sensor networks", *IEEE Transactions on Parallel and Distributed Systems*, vol. 26, no. 7, pp. 1971–1983, July 2015.

[LIN 05] LIN F. Y.S., CHIU P.L., "A near-optimal sensor placement algorithm to achieve complete coverage/discrimination in sensor networks", *IEEE Communications Letters*, vol. 9, no. 1, pp. 43–45, 2005.

[LIU 12] ZHONG LIU J., LEI WANG B., JUN-YU A. *et al.*, "An immune-swarm intelligence based algorithm for deterministic coverage problems of wireless sensor networks", *J. Cent. South Univ.*, vol. 19, pp. 3154–3161, 2012.

[LIU 14] LIU X., HE D., "Ant colony optimization with greedy migration mechanism for node deployment in wireless sensor networks", *Journal of Network and Computer Applications*, vol. 39, pp. 310–318, 2014.

[LOR 04] LORINCZ K., MALAN D., FULFORD-JONES T. *et al.*, "Sensor networks for emergency response: challenges and opportunities", *IEEE Pervasive Computing*, vol. 3, no. 4, pp. 16–23, 2004.

[LED 05] LÈDECZI A., NĀDAS A., VÖLGYESI P. *et al.*, "Countersniper system for urban warfare", *ACM Transactions on Sensor Networks*, vol. 1, no. 2, pp. 153–177, November 2005.

[MAH 13a] MAHBOUBI H., HABIBI J., AGHDAM A.G. *et al.*, "Distributed deployment strategies for improved coverage in a network of mobile sensors with prioritized sensing field", *IEEE Transactions on Industrial Informatics*, vol. 9, no. 1, pp. 451–461, February 2013.

[MAH 13b] MAHFOUDH S., KHOUFI I., MINET P. *et al.*, "Relocation of mobile wireless sensors in the presence of obstacles", *20th International Conference on Telecommunications (ICT)*, Casablanca, Morocco, pp. 1–5, May 2013.

[MAH 14a] MAHBOUBI H., MOEZZI K., AGHDAM A.G. *et al.*, "Distributed deployment algorithms for efficient coverage in a network of mobile sensors with nonidentical sensing capabilities", *IEEE Transactions on Vehicular Technology*, vol. 63, no. 8, pp. 3998–4016, October 2014.

[MAH 14b] MAHBOUBI H., MOEZZI K., AGHDAM A.G. *et al.*, "Distributed deployment algorithms for improved coverage in a network of wireless mobile sensors", *IEEE Transactions on Industrial Informatics*, vol. 10, no. 1, pp. 163–174, 2014.

[MEG 01] MEGUERDICHIAN S., KOUSHANFAR F., POTKONJAK M. *et al.*, "Coverage problems in wireless ad-hoc sensor networks", *Proceedings of IEEE INFOCOM*, vol. 3, Anchorage, AK, pp. 1380–1387, 2001.

[MEG 02] MEGERIAN S., KOUSHANFAR F., QU G. *et al.*, "Exposure in wireless sensor networks: Theory and practical solutions", *Journal of Wireless Networks*, vol. 8, no. 5, pp. 443–454, 2002.

[MIA 15] MIAO C., DAI G., ZHAO X. *et al.*, "3D self-deployment algorithm in mobile wireless sensor networks", *International Journal of Distributed Sensor Networks*, vol. 2015, p. 11, October 2015.

[MIN 13] MINI S., UDGATA S.K., SABAT S.L., "Artificial bee colony algorithm for probabilistic target Q-coverage in wireless sensor networks", in PANIGRAHI B.K. *et al.* (eds), *SEMCCO Part I: LNCS 8297*, Springer International Publishing Switzerland, 2013.

[MO 09] MO L., HE Y., LIU Y. *et al.*, "Canopy closure estimates with greenorbs: Sustainable sensing in the forest", *Proceedings of the 7th ACM Conference on Embedded Networked Sensor Systems (SenSys'09)*, New York, ACM, pp. 99–112, 2009.

[MOO 10] MOO CROSSBOW, "Crossbow Imote2 Mote Specifications", available at http://www.xbow.com/, 2010.

[NAG 15] NAGCHOUDHURY P., MAHESHWARI S., CHOUDHARY K., "Optimal sensor nodes deployment method using bacteria foraging algorithm in wireless sensor networks", *Emerging ICT for Bridging the Future – Volume 2*, of *Advances in Intelligent Systems and Computing*, Springer International Publishing, vol. 338 pp. 221–228, 2015.

[NAK 07] NAKAMURA E.F., LOUREIRO A.F., FRERY A.C., "Information fusion for wireless sensor networks: methods, models, and classifications", *ACM Comput. Surv.*, vol. 39, no. 3, p. 55, August 2007.

[NAZ 14] NAZI A., RAJ M., FRANCESCO M.D. *et al.*, "Deployment of robust wireless sensor networks using gene regulatory networks: An isomorphism-based approach", *Pervasive and Mobile Computing*, vol. 13, pp. 246–257, 2014.

[NI 08] NI Y., ZHOU H., CHAN K. *et al.*, "Modal flexibility analysis of cable-stayed Ting Kau Bridge for damage identification", *Computer-Aided Civil and Infrastructure Eng.*, vol. 23, no. 3, pp. 223–236, 2008.

[NIK 93] NIKOOKAR H., HASHEMI H., "Statistical modeling of signal amplitude fading of indoor radio propagation channels", *Personal Communications: Gateway to the 21st Century*, Ottawa, vol. 1, pp. 84–88, 1993.

[O'DO 13] O'DONOVAN T., BROWN J., BÜSCHING F. *et al.*, "The GINSENG system for wireless monitoring and control: design and deployment experiences", *ACM Trans. Sensor Netw.*, vol. 10, no. 1, p. 40, November 2013.

[ONU 07] ONUR E., ERSOY C., DELIC H. *et al.*, "Surveillance wireless sensor networks: deployment quality analysis", *IEEE Network*, vol. 21, no. 6, pp. 48–53, November-December 2007.

[ORO 87] OROURKE J., *Art Gallery Theorems and Algorithms*, Oxford University Press, 1987.

[OSM 11] OSMANI A., "Design and evaluation of two distributed methods for sensors placement in wireless sensor networks", *Journal of Advances in Computer Research*, vol. 2, no. 1, pp. 13–26, February 2011.

[OZT 11] OZTURK C., KARABOGA D., GORKEMLI B., "Probabilistic dynamic deployment of wireless sensor networks by artificial bee colony algorithm", *Sensors*, vol. 11, no. 6, pp. 6056–6065, 2011.

[PAP 81] PAPADIMITRIOU C.H., "On the complexity of integer programming", *J. ACM*, vol. 28, no. 4, pp. 765–768, ACM, 1981.

[PEN 03] PENROSE M., *Random Geometric Graphs*, vol. 5, Oxford University Press, 2003.

[PEN 08] PENG L., WANG D., ZHAO L., "An algorithm based on market competition for wireless sensor network connectivity and coverage", *International Conference on Information and Automation (ICIA)*, Changsha, pp. 379–383, 2008.

[PET 06] PETROVA M., RIIHIJARVI J., MAHONEN P. *et al.*, "Performance study of IEEE 802.15.4 using measurements and simulations", *IEEE Wireless Communications and Networking Conference (WCNC'06)*, vol. 1, Las Vegas, NV, pp. 487–492, 2006.

[POD 04] PODURI S., SUKHATME G., "Constrained coverage for mobile sensor networks", *Proceedings of the IEEE International Conference on Robotics and Automation (ICRA'04)*, vol. 1, pp. 165–171, 2004.

[POM 06] POMPILI D., MELODIA T., AKYILDIZ I.F., "Deployment analysis in underwater acoustic wireless sensor networks", *Proceedings of the ACM International Workshop on UnderWater Networks (WUWNet)*, Los Angeles, CA, pp. 48–55, September 2006.

[QI 08] QI G., SONG P., LI K., "Blackboard mechanism based ant colony theory for dynamic deployment of mobile sensor networks", *Journal of Bionic Engineering*, vol. 5, no. 3, pp. 197–203, 2008.

[RAP 01] RAPPAPORT T., *Wireless Communications: Principles and Practice*, 2nd ed., Prentice Hall, 2001.

[RAZ 13] RAZZAQUE M.A., BLEAKLEY C., DOBSON S., "Compression in wireless sensor networks: a survey and comparative evaluation", *ACM Trans. Sensor Netw.*, vol. 10, no. 1, p. 44, November 2013.

[REB 15] REBAI M., BERRE M.L., SNOUSSI H. *et al.*, "Sensor deployment optimization methods to achieve both coverage and connectivity in wireless sensor networks", *Computers & Operations Research*, vol. 59, pp. 11–21, 2015.

[SAH 08] SAHIN C.S., URREA E., UYAR M.U. *et al.*, "Self-deployment of mobile agents in MANETs for military applications", *Army Science Conference*, pp. 1–8, 2008.

[SEN 11] SENOUCI M., MELLOUK A., OUKHELLOU L. *et al.*, "Uncertainty-aware sensor network deployment", *IEEE Global Telecommunications Conf. (GLOBECOM'11)*, Houston, Texas, USA, pp. 1–5, December 2011.

[SEN 12a] SENOUCI M., MELLOUK A., AISSANI A., "An analysis of intrinsic properties of stochastic node placement in sensor networks", *IEEE Global Telecommunications Conf. GLOBECOM'12*, Houston, Texas, USA, pp. 494–499, December 2012.

[SEN 12b] SENOUCI M., MELLOUK A., OUKHELLOU L. *et al.*, "Efficient uncertainty-aware deployment algorithms for wireless sensor networks", *IEEE Wireless Communications and Networking Conf. (WCNC'2012)*, Paris, France, pp. 2163–2167, April 2012.

[SEN 12c] SENOUCI M., MELLOUK A., OUKHELLOU L. *et al.*, "An evidence-based sensor coverage model", *IEEE Communications Letters*, vol. 16, no. 9, pp. 1462–1465, 2012.

[SEN 12d] SENOUCI M., MELLOUK A., OUKHELLOU L. *et al.*, "Using the belief functions theory to deploy static wireless sensor networks", in DENOEUX T., MASSON M.-H. (eds), *Belief Functions: Theory and Applications*, Springer Berlin/Heidelberg, 2012.

[SEN 13a] SENEL F., AKKAYA K., YILMAZ T., "Autonomous deployment of sensors for maximized coverage and guaranteed connectivity in underwater acoustic sensor networks", *38th IEEE Conference on Local Computer Networks (LCN)*, Sydney, Australia, pp. 211–218, October 2013.

[SEN 13b] SENOUCI M., ABDELHAMID M., ASSNOUNE K., "Localized movement-assisted sensor deployment algorithm for hole detection and healing", *IEEE Transactions on Parallel and Distributed Systems*, vol. 25, no. 5, pp. 1267–1277, 2013.

[SEN 14a] SENOUCI M., On the use of the belief functions theory in the deployment and control of wireless sensor networks, PhD Thesis, USTHB-Algeria/UPEC-France, January 2014.

[SEN 14b] SENOUCI M., BOUDAREN M., SENOUCI M.A. *et al.*, "A smart methodology for deterministic deployment of wireless sensor networks", *IEEE International Conference on Smart Communications in Network Technologies (SaCoNeT)*, pp. 103–108, 18–20 June 2014.

[SEN 14c] SENOUCI M., BOUGUETTOUCHE D., SOUILAH F. *et al.*, "Efficient heuristic for the deterministic deployment of wireless sensor networks", *The International Conference on Metaheuristics and Nature Inspired Computing*, Marrakech, Morocco, 2014.

[SEN 14d] SENOUCI M., SOUILAH F., BOUGUETTOUCHE D. *et al.*, "Simulated annealing for solving the wireless sensor networks deployment problem", *The International Conference on Metaheuristics and Nature Inspired Computing*, Marrakech, Morocco, 2014.

[SEN 14e] SENOUCI M., MELLOUK A., AISSANI A., "Random deployment of wireless sensor networks: a survey and approach", *Int. J. Ad Hoc and Ubiquitous Computing*, vol. 15, no. 1/2/3, pp. 133–146, 2014.

[SEN 15a] SENOUCI M., BOUGUETTOUCHE D., SOUILAH F. *et al.*, "Static wireless sensor networks deployment using an improved binary PSO", *International Journal of Communication Systems*, vol. PP, pp. 1–17, 2015.

[SEN 15b] SENOUCI M., MELLOUK A., ASNOUNE K. *et al.*, "Movement-assisted sensor deployment algorithms: a survey and taxonomy", *IEEE Communication Surveys & Tutorials*, vol. 17, no. 4, pp. 2493–2510, March 2015.

[SEN 15c] SENOUCI M., MELLOUK A., OUKHELLOU L. *et al.*, "WSNs deployment framework based on the theory of belief functions", *Journal of Computer Networks*, vol. 88, pp. 12–26, 2015.

[SER 10] SERGIOU C., VASSILIOU V., "Energy utilization of HTAP under specific node placements in Wireless Sensor Networks", *European Wireless Conference (EW)*, Lucca, Italy, pp. 482–487, April 2010.

[SEV 14] SEVGI C., KOÇYIGIT A., "Optimal deployment in randomly deployed heterogeneous WSNs: A connected coverage approach", *Journal of Network and Computer Applications*, vol. 46, pp. 182–197, 2014.

[SHU 05] SHU H., LIANG Q., "Fuzzy optimization for distributed sensor deployment", *IEEE Wireless Communications and Networking Conference*, vol. 3, pp. 1903–1908, 2005.

[SIB 02] SIBLEY G., RAHIMI M., SUKHATME G., "Robomote: a tiny mobile robot platform for large-scale ad-hoc sensor networks", *Proceedings of the IEEE International Conference on Robotics and Automation (ICRA'02)*, vol. 2, Washington, DC, pp. 1143–1148, 2002.

[SIL 14] SILVA R., SILVA J.S., BOAVIDA F., "Mobility in wireless sensor networks – Survey and proposal", *Computer Communications*, vol. 52, pp. 1–20, 2014.

[SIN 07] SINHA A., PAL B., "Stensor: a novel stochastic algorithm for placement of sensors in a rectangular grid", *Annual Techno-Management Fest*, 2007.

[SUE 08] SUEN Y., A genetic-algorithm based mobile sensor network deployment algorithm, Report, Department of Electrical and Computer Engineering, University of Texas, 2008.

[SUG 11] SUGIHARA R., GUPTA R.K., "Path planning of data mules in sensor networks", *ACM Trans. Sen. Netw.*, ACM, vol. 8, no. 1, pp. 1–27, 2011.

[TAM 10] TAMBOLI N., YOUNIS M., "Coverage-aware connectivity restoration in mobile sensor networks", *Journal of Network and Computer Applications*, vol. 33, pp. 363–374, 2010.

[TAN 09] TAN G., JARVIS S.A., KERMARREC A.M., "Connectivity-guaranteed and obstacle-adaptive deployment schemes for mobile sensor networks", *IEEE Transactions on Mobile Computing*, vol. 8, no. 6, pp. 836–848, 2009.

[TAN 11] TAN R., XING G., LIU B. *et al.*, "Exploiting data fusion to improve the coverage of wireless sensor networks", *IEEE/ACM Transactions on Networking*, vol. 20, no. 2, pp. 450–462, 2011.

[TEM 14] TEMEL S., UNALDI N., KAYNAK O., "On deployment of wireless sensors on 3-D terrains to maximize sensing coverage by utilizing cat swarm optimization with wavelet transform", *IEEE Transactions on Systems, Man, and Cybernetics: Systems*, vol. 44, no. 1, pp. 111–120, January 2014.

[TIA 05] TIAN D., GEORGANAS N.D., "Connectivity maintenance and coverage preservation in wireless sensor networks", *Ad Hoc Networks*, vol. 3, no. 6, pp. 744–761, Elsevier Science Publishers B.V., 2005.

[TOP 11] TOPCUOGLU H., ERMIS M., SIFYAN M., "Positioning and utilizing sensors on a 3-D terrain Part I-Theory and modeling", *IEEE Trans. Syst., Man, Cybern. C, Appl. Rev*, vol. 41, no. 3, pp. 376–382, May 2011.

[VAL 13] VALES-ALONSO J., PARRADO-GARCIA F., LOPEZ-MATENCIO P. *et al.*, "On the optimal random deployment of wireless sensor networks in non-homogeneous scenarios", *Ad Hoc Networks*, vol. 11, pp. 846–860, 2013.

[VAR 96] VARSHNEY P., *Distributed Detection and Data Fusion*, Springer-Verlag, 1996.

[VAS 09] VASSILIOU V., SERGIOU C., "Performance study of node placement for congestion control in wireless sensor networks", *3rd International Conference on New Technologies, Mobility and Security (NTMS)*, Cairo, pp. 1–8, December 2009.

[VIC 09] VICAIRE P., HE T., CAO Q. *et al.*, "Achieving long-term surveillance in VigilNet", *ACM Trans. Sensor Netw.*, vol. 5, no. 1, p. 39, February 2009.

[VIE 04] VIEIRA L., VIEIRA M., BEATRIZ L. *et al.*, "Efficient incremental sensor network deployment algorithm", *Brazilian Symposium on Computer Networks*, 2004.

[WAD 09] WADHWA M., SONG M., RALI V. *et al.*, "The impact of antenna orientation on wireless sensor network performance", *2nd IEEE International Conference on Computer Science and Information Technology (ICCSIT'09)*, Beijing, pp. 143–147, 2009.

[WAH 07] WAH B.W., CHEN Y., WANG T., "Simulated annealing with asymptotic convergence for nonlinear constrained optimization", *J. of Global Optimization*, Kluwer Academic Publishers, vol. 39, no. 1, pp. 1–37, 2007.

[WAN 03] WANG X., XING G., ZHANG Y. *et al.*, "Integrated coverage and connectivity configuration in wireless sensor networks", *Proceedings of the 1st international ACM conference on Embedded networked sensor systems (SenSys'03)*, New York, USA, pp. 28–39, 2003.

[WAN 05a] WANG G., CAO G., LA PORTA T. *et al.*, "Sensor relocation in mobile sensor networks", *Proceedings of the 24th Annual Joint Conference of the IEEE Computer and Communications Societies (INFOCOM)*, vol. 4, Miami, FL, USA, pp. 2302–2312, 2005.

[WAN 05b] WANG Y.-C., HU C.-C., TSENG Y.-C., "Efficient deployment algorithms for ensuring coverage and connectivity of wireless sensor networks", *Proc. First Int Wireless Internet Conf*, pp. 114–121, 2005.

[WAN 06a] WAN P., YI C., "Coverage by randomly deployed wireless sensor networks", *IEEE/ACM Trans. Netw.*, vol. 14, no. SI, pp. 2658–2669, IEEE Press, 2006.

[WAN 06b] WANG G., CAO G., PORTA T.L., "Movement-assisted sensor deployment", *IEEE Transactions on Mobile Computing*, vol. 5, no. 6, pp. 640–652, 2006.

[WAN 06c] WANG P.C., HOU T.W., YAN R.H., "Maintaining coverage by progressive crystal-lattice permutation in mobile wireless sensor networks", *Proceedings of the International Conference on Systems and Networks Communication*, Tahiti, p. 42, October 2006.

[WAN 06d] WANG X., WANG S., MA J., "Dynamic deployment optimization in wireless sensor networks", HUANG D.-S., LI K., IRWIN G. (ed.), *Intelligent Control and Automation*, Springer Berlin Heidelberg, 2006.

[WAN 06e] WANG X., YANG Y., SONG Y., "Redundant movement-assisted sensor deployment based on virtual rhomb grid in wireless sensor networks", *Proceedings of the IEEE International Conference on Mechatronics and Automation*, Luoyang, Henan, pp. 775–779, 2006.

[WAN 07a] WANG G.G., CAO G., BERMAN P. *et al.*, "Bidding protocols for deploying mobile sensors", *IEEE Transactions on Mobile Computing*, vol. 6, no. 5, pp. 563–576, 2007.

[WAN 07b] WANG W., SRINIVASAN V., CHUA K. *et al.*, "Energy-efficient coverage for target detection in wireless sensor networks", *6th International Conference on Information Processing in Sensor Networks (IPSN)*, Cambridge, MA, pp. 313–322, 2007.

[WAN 08a] WANG D., XIE B., AGRAWAL D., "Coverage and lifetime optimization of wireless sensor networks with Gaussian distribution", *IEEE Transactions on Mobile Computing*, vol. 7, no. 12, pp. 1444–1458, 2008.

[WAN 08b] WANG Y.-C., TSENG Y.-C., "Distributed deployment schemes for mobile wireless sensor networks to ensure multilevel coverage", *IEEE Transactions on Parallel and Distributed Systems*, vol. 19, no. 9, pp. 1280–1294, 2008.

[WAN 09] WANG B., LIM H.B., MA D., "A survey of movement strategies for improving network coverage in wireless sensor networks", *Computer Communications*, vol. 32, nos. 13–14, pp. 1427–1436, 2009.

[WAN 10a] WANG B., *Coverage Control in Sensor Networks*, Springer, New York, 2010.

[WAN 10b] WANG C.-F., LEE C.-C., "The optimization of sensor relocation in wireless mobile sensor networks", *Computer Communications*, vol. 33, pp. 828–840, 2010.

[WAN 11] WANG B., "Coverage problems in sensor networks: a survey", *ACM Computing Surveys*, vol. 43, no. 4, p. 53, October 2011.

[WER 06] WERNER-ALLEN G., LORINCZ K., RUIZ M. *et al.*, "Deploying a wireless sensor network on an active volcano", *IEEE Internet Computing*, vol. 10, no. 2, pp. 18–25, March/April 2006.

[WON 04] WONG T., TSUCHIYA T., KIKUNO T., "A self-organizing technique for sensor placement in wireless micro-sensor networks", *18th International Conference on Advanced Information Networking and Applications (AINA 2004)*, vol. 1, pp. 78–83, 2004.

[WON 14] Won M., Ra H., Park T. *et al.*, "Modeling random deployment in wireless sensor networks for infrastructure less cyber physical systems", *2nd IEEE International Conference on Cyber-Physical Systems, Networks, and Applications*, Hong Kong, pp. 81–86, 2014.

[WOO 08] Wood A., Stankovic J., Virone G. *et al.*, "Context-aware wireless sensor networks for assisted living and residential monitoring", *IEEE Network*, vol. 22, no. 4, pp. 26–33, 2008.

[WU 07a] Wu C.-H., Lee K.-C., Chung Y.-C., "A Delaunay triangulation based method for wireless sensor network deployment", *Comput. Commun.*, vol. 30, nos. 14–15, pp. 2744–2752, Elsevier Science Publishers B.V., 2007.

[WU 07b] Wu J., Yang S., "Optimal movement-assisted sensor deployment and its extensions in wireless sensor networks", *Simulation Modelling Practice and Theory*, vol. 15, no. 4, pp. 383–399, 2007.

[XIA 14] Xiao F., Yang Y., Wang R. *et al.*, "A novel deployment scheme based on three-dimensional coverage model for wireless sensor networks", *The Scientific World Journal*, vol. 2014, p. 7, 2014.

[XIN 05] Xing G., Wang X., Zhang Y. *et al.*, "Integrated coverage and connectivity configuration for energy conservation in sensor networks", *ACM Transactions on Sensor Networks*, vol. 1, no. 1, pp. 36–72, August 2005.

[XIN 09] Xing G., Tan R., Liu B. *et al.*, "Data fusion improves the coverage of wireless sensor networks", *Proceedings of the 15th Annual International Conference on Mobile Computing and Networking (MobiCom'09)*, Beijing, China, pp. 157–168, 2009.

[XU 06] Xu Y., Yao X., "A GA approach to the optimal placement of sensors in wireless sensor networks with obstacles and preferences", *Proceedings of the IEEE Conference on Consumer Communications and Networking*, January 2006.

[YAN 06a] Yang G., *Body Sensor Networks*, Springer, Berlin, 2006.

[YAN 06b] Yang S., Wu J., Dai F., "Localized movement-assisted sensor deployment in wireless sensor networks", *IEEE International Conference on Mobile Adhoc and Sensor Systems (MASS)*, Vancouver, BC, pp. 753–758, 2006.

[YAN 07] Yang S., Li M., Wu J., "Scan-based movement-assisted sensor deployment methods in wireless sensor networks", *IEEE Transactions on Parallel and Distributed Systems*, vol. 18, no. 8, pp. 1108–1121, 2007.

[YAN 08a] Yan T., Gu Y., He T. *et al.*, "Design and optimization of distributed sensing coverage in wireless sensor networks", *ACM Trans. Embedd. Comput. Syst.*, vol. 7, no. 3, p. 33, April 2008.

[YAN 08b] Yang M., Cao Y., Tan L. *et al.*, "An enhanced self-deployment algorithm in mobile sensor network", *International Seminar on Future Information Technology and Management Engineering (FITME'08)*, Leicestershire, United Kingdom, pp. 573–576, 2008.

[YAN 08c] Yang M., Cao Y., Tan L. *et al.*, "A new self-deployment algorithm in hybrid sensor network", *Second International Symposium on Intelligent Information Technology Application (IITA'08)*, vol. 2, Shanghai, pp. 268–272, December 2008.

[YAN 09] YANG G., QIAO D., "Barrier information coverage with wireless sensors", *IEEE INFOCOM*, pp. 918–926, 2009.

[YON 09] YONG Z., LI W., "A sensor deployment algorithm for mobile wireless sensor networks", *Proceedings of the 2nd Annual International Conference on Chinese Control and Decision Conference*, Guilin, China, pp. 4642–4647, 2009.

[YOO 08] YOO Y., AGRAWAL P., "Mobile sensor relocation to prolong the lifetime of wireless sensor networks", *IEEE Vehicular Technology Conference*, Singapore, pp. 193–197, May 2008.

[YOU 08] YOUNIS M., AKKAYA K., "Strategies and techniques for node placement in wireless sensor networks: a survey", *Ad Hoc Netw.*, vol. 6, pp. 621–655, Elsevier Science Publishers B.V., June 2008.

[YOU 10] YOUNIS M., LEE S., ABBASI A.A., "A localized algorithm for restoring internode connectivity in networks of moveable sensors", *IEEE Transactions on Computers*, vol. 59, no. 12, pp. 1669–1682, December 2010.

[YOU 14] YOUNIS M., SENTURK I.F., AKKAYA K. *et al.*, "Topology management techniques for tolerating node failures in wireless sensor networks: A survey", *Computer Networks*, vol. 58, pp. 254–283, 2014.

[YU 13a] YU X., LIU N., HUANG W. *et al.*, "A node deployment algorithm based on Van Der Waals force in wireless sensor networks", *International Journal of Distributed Sensor Networks*, vol. 2013, p. 8, 2013.

[YU 13b] YU X., ZHANG J., FAN J. *et al.*, "A faster convergence artificial bee colony algorithm in sensor deployment for wireless sensor networks", *International Journal of Distributed Sensor Networks*, vol. 2013, p. 9, 2013.

[YUA 08] YUAN Z., TAN R., XING G. *et al.*, "Fast sensor placement algorithms for fusion-based target detection", *Real-Time Systems Symposium*, Barcelona, pp. 103–112, 2008.

[YUN 10] YUN Z., BAI X., XUAN D. *et al.*, "Optimal deployment patterns for full coverage and k-connectivity ($k \leq 6$) wireless sensor networks", *IEEE/ACM Trans. Netw.*, vol. 18, no. 3, pp. 934–947, 2010.

[ZHA 04a] ZHANG H., HOU J., "On deriving the upper bound of α-lifetime for large sensor networks", *Proceedings of the 5th ACM International Symposium on Mobile Ad Hoc Networking and Computing (MobiHoc'04)*, Roppongi Hills, Tokyo, Japan, pp. 121–132, 2004.

[ZHA 04b] ZHANG P., SADLER C., LYON S. *et al.*, "Hardware design experiences in ZebraNet", *In Proceedings of ACM SenSys'04*, Baltimore, MD, USA, November 2004.

[ZHA 05] ZHANG H., HOU J., "Maintaining sensing coverage and connectivity in large sensor networks", *Ad Hoc & Sensor Wireless Networks*, vol. 1, nos. 1–2, pp. 89–124, 2005.

[ZHA 06] ZHANG J., YAN T., SON S.H., "Deployment strategies for differentiated detection in wireless sensor networks", *3rd Annual IEEE Communications Society on Sensor and Ad Hoc Communications and Networks (SECON)*, vol. 1, pp. 316–325, 2006.

[ZHA 10] ZHANG C., BAI X., TENG J. *et al.*, "Constructing low-connectivity and full-coverage three dimensional sensor networks", *IEEE Journal on Selected Areas in Communications*, vol. 28, no. 7, pp. 984–993, September 2010.

[ZHO 06] ZHOU G., HE T., KRISHNAMURTHY S. *et al.*, "Models and solutions for radio irregularity in wireless sensor networks", *ACM Transactions on Sensor Networks*, vol. 2, no. 2, pp. 221–262, 2006.

[ZHU 15] ZHU J., WANG B., "The optimal placement pattern for confident information coverage in wireless sensor networks", *IEEE Transactions on Mobile Computing*, no. 99, pp. 1–11, 2015.

[ZOU 03a] ZOU Y., CHAKRABARTY K., "Sensor deployment and target localization based on virtual forces", *Twenty-Second Annual Joint Conference of the IEEE Computer and Communications (INFOCOM)*, vol. 2, San Francisco, CA, pp. 1293–1303, 2003.

[ZOU 03b] ZOU Y., CHAKRABARTY K., "Uncertainty-aware sensor deployment algorithms for surveillance applications", *Proc. IEEE Global Telecommunications Conf. (GLOBECOM'03)*, vol. 5, pp. 2972–2976, 2003.

[ZOU 05] ZOU Y., CHAKRABARTY K., "A distributed coverage- and connectivity-centric technique for selecting active nodes in wireless sensor networks", *IEEE Transactions on Computers*, vol. 54, no. 8, pp. 978–991, 2005.

Index

H, I, K, L

M, N, O

P, Q, R, S

T, V, W, X

Printed in the United States
By Bookmasters